Applied Ergonomics Handbook

Mike Burke

Library of Congress Cataloging-in-Publication Data

Burke, Mike, 1953–
Applied ergonomics handbook / Mike Burke.
p. cm.
Includes index.
ISBN 0-87371-367-2
1. Human engineering—Handbooks, manuals, etc. I. Title.
TA166.B89 1992
620.8′2—dc20 91-34532
CIP

The names of individuals or companies used in this book are fictitious. Any resemblance to actual persons or organizations is coincidental.

Direct all inquiries to CRC Press, Inc., 2000 Corporate Blvd., N.W., Boca Raton, Florida 33431.

PRINTED IN THE UNITED STATES OF AMERICA
2 3 4 5 6 7 8 9 0
Printed on acid-free paper

To my wife Teri
and my sons
Luke and Adam

AUTHOR

Mike Burke is the Manager of Ergonomics Programs for the Physical Therapy Division of Baxter Healthcare Corporation. In that capacity, he works as a consultant with organizations and companies to develop and implement various cumulative trauma injury reduction strategies.

He has provided services in over 100 different clerical, manufacturing, and distribution settings over the past five years.

Mr. Burke earned his degree in physical therapy from Northwestern University Medical School in 1985. Prior to becoming a physical therapist, he worked in different types of industry developing a diverse background helpful in analyzing specific ergonomic challenges.

He has written several articles and created an ergonomic analysis software program using the procedures described in this book. He has developed several unique ergonomics educational programs to help instruct persons at various levels in the basic and simple application of ergonomic principles. He has also given entertaining and lively ergonomics presentations at many national professional meetings.

PREFACE

This book is for all the occasional ergonomists. An occasional ergonomist is one who has several other duties to perform in addition to playing a role in controlling musculoskeletal cumulative trauma. You could be a health and safety professional, an on-site health care provider, a member of an ergonomics committee, a buyer, or even a health care provider wishing to deliver an additional service.

INTRODUCTION

Ergonomics is the science that facilitates maximum productivity, consistent quality, and long-term worker health and safety.

Of these three goals the most challenging is the long-term health and safety of the worker. The aspect of health and safety takes into account a spectrum of injury and illness prevention goals. One of the most challenging is the control of cumulative trauma disorders. This book will deal with the issue of controlling cumulative trauma disorders by focusing on the musculoskeletal mechanical stresses in the workplace. It should be understood that while your efforts may be driven by the need to control cumulative trauma disorders, maximum productivity and consistent quality will surely result.

The science of ergonomics as an injury-reduction strategy is evolving. A part of that evolution is to try to arrive at an objective system to measure ergonomic risk. This system will need to be as objective as possible, while at the same time it will need to be flexible enough to consider all job situations. It should provide consistent and reproducible results and at the same time be practical to implement.

This book will describe a model for determining ergonomic needs in specific areas, performing the analyses of those targeted areas, discovering various interventions to decrease or eliminate potentially hazardous conditions, screening those interventions, and implementing them.

The purpose of the analysis model described in this book is to provide a logical basis for initiating any program or modification to a workstation. Every company today is faced with a mass of options. Can a problem be solved with an education program as simple as a poster, or is more significant workstation redesign necessary?

Considering all the different approaches and devices available, it is apparent that no one has discovered a single answer to the problem. Most of the real success has come as a result of a comprehensive approach. It has included many different approaches but has had at its core a strong commitment on the part of management. Whether it was the plant manager, division director, or C.E.O., someone has taken the stand that "safety" was now an important part of quality and profit. In many companies, management has realized that to continue to ignore this problem is going to be more expensive than initiating a genuine action plan. That plan will involve communication, implementation, and accountability.

There is no clear-cut, totally objective, and universally accepted method for identifying ergonomic risk factors or determining the most appropriate action plan. This is because primarily what you are dealing with is *cumulative trauma*. Cumulative trauma is an injury or illness that is the result of multiple stresses or exposures to stresses over time. The actual illness or injury may be a direct result of the stress or the human body's adaptation to these stresses. An example is when certain muscles overdevelop and others become tight or weak. In time this adaptation to a "normal day's" work load may develop into an injury, such as a low-back injury. Another possible cause is when a body part is repetitively injured, often on a cellular level. In this case, the body does not have a chance to adapt. In most cases, it is a combination of factors that leads to a cumulative trauma disorder.

If you are dealing with a sudden burn, or a foreign body in the eye, it is simple to identify the cause. A person reaches out and touches a red-hot piece of metal, or a piece of flying debris bypasses safety glasses. It results in an acute and immediate injury. When it comes to cumulative trauma, the relationship between the cause and the resulting injury or illness is not as easily defined.

If a person develops shoulder tendonitis, it is difficult, if not impossible, to absolutely determine the cause. While historical data can identify certain factors that have been associated with the development of tendonitis, you can never be absolutely certain of the cause of a cumulative injury or illness.

In addition, there is virtually no factual information relative to the dose response. The dose response is the number of times or amount of time a person must be exposed to a particular stress before he or she develops an injury or illness. In the case of a burn or a foreign body in the eye, the dose response is 1. A worker need only be exposed once to be injured. In the case of cumulative trauma, it is not known how many times a person can forward bend at the waist and lift incorrectly before they will suffer a back injury, or how many times a person can be exposed to forceful exertion with an awkward wrist posture and vibration before developing carpal tunnel syndrome.

Many ergonomic risk factors can be identified based on research on the incidence of injury, medical histories, biomechanics, anatomy, and physiology. The challenge is to determine the most effective strategy for controlling or eliminating the potentially harmful effects.

This book takes the approach of a how-to guide for the performance of ergonomics analyses. It is the result of years of job-site analysis experience using various techniques to find the procedures that are the least cumbersome, most practical, and most objective. Take from the procedures, forms, and advice given here what you find appropriate for your work, and modify whatever does not fit into your personal or professional philosophy.

Anyone with a sincere interest in learning about this field will be able to develop the skills to identify and solve most of the conditions likely to be present in a workstation. This may include a health and safety professional, a member of an ergonomics committee, or an occupational health care provider who has a limited amount of time to focus on ergonomics analysis.

This book provides the necessary tools to generate a report that comprehensively describes specific risk factors and suggests interventions to control those risk factors. Also included are guidelines for screening and implementing interventions, ergonomic committee formation, and the establishment of a corporate- or company-wide ergonomics process.

The ergonomic process described in this book can be broken into seven general steps:

1. Determining measurement criteria and target jobs
2. Gathering job background information
3. Identifying ergonomic risk factors
4. Discovering interventions
5. Screening interventions
6. Implementing interventions
7. Tracking the effectiveness of those interventions

These steps can be summarized as follows:

DETERMINING MEASUREMENT CRITERIA AND TARGET JOBS

Determining measurement criteria involves determining which areas in a facility should be targeted for analysis and intervention. Specific criteria, such as injured days off or injury incidence rates, are selected and utilized to make this decision. In many cases, additional or completely different criteria will be used to measure the effectiveness of an intervention. These alternate or additional criteria may be turnover rates, productivity variances, or subjective comfort-level survey results.

GATHERING JOB BACKGROUND INFORMATION

Comprehensive job background information needs to be collected on each targeted area. This includes the identification of the input and output of the operation; a narrative description of the job; and a breakdown of duties, tasks, and, in some cases, steps and components of the job. Input is gathered from human resources, management, and hourly workers. Objective measurement of workstation distances and heights, and static and dynamic forces, is also performed. The targeted area is observed, and in many cases, videotaped. Additional information is gathered from the worker whenever appropriate.

IDENTIFYING ERGONOMIC RISK FACTORS

Using an anatomical region approach, the postural or position stresses, force exposure stresses, environmental factors, and energy expenditure considerations are identified and listed. Each risk factor is associated with the corresponding duty, task, or step of the job. The result is a comprehensive list of ergonomic risk factors.

DISCOVERING INTERVENTIONS

Each risk factor is considered for reduction utilizing four different approaches. Changes in the input or materials, changes in the output or product, changes in the machine or environment, or actions dealing with the worker are considered and listed. Using a brainstorming approach, a list of interventions, from the most practical to the most farfetched, is compiled.

SCREENING INTERVENTIONS

A specific approach is determined and each intervention is considered for implementation. Factors such as cost, return on investment, projected effectiveness, and relative effects on productivity and performance are considered. The most appropriate intervention is chosen.

IMPLEMENTING INTERVENTIONS

The first step is to let the individuals who are going to be affected by this intervention know that a decision has been made. The next step is to orient that group, being sure to cover why this particular intervention was chosen, why the individuals were chosen, what the expected impact is, and whom they can contact should any unforeseen condition result. The goal is to maximize compliance, and, ideally, to foster enthusiasm for the project.

TRACKING THE EFFECTIVENESS OF THOSE INTERVENTIONS

The criteria for the tracking of effectiveness was already selected in the first step. The last step will be to assess the effectiveness and to decide whether to expand, amend, alter, or abandon a particular intervention.

CONTENTS

THE ERGONOMICS PROCESS

START ERGONOMICS PROCESS	COMPANY LOGISTICS BACKGROUND	SELECT THE CRITERIA	COMPUTE TOTAL INCIDENCE	IDENTIFY PROBLEM AREAS	SELECT TARGET JOBS	SELECT RECORDING PROCEDURES	COMFORT-LEVEL SURVEY
PREPARE MANAGEMENT REPORT							START ERGONOMICS COMMITTEE
MEASURE SOLUTION SUCCESS							COMPILE JOB DESCRIPTION
IMPLEMENT EFFECTIVE SOLUTIONS							DETERMINE JOB EXPOSURE
CHOOSE EFFECTIVE SOLUTIONS							LIST OF DUTIES
FIND ERGONOMIC SOLUTIONS							INTERVIEW SELECTED WORKERS
QUANTIFY RISK FACTORS							DETERMINE DUTY EXPOSURE
IDENTIFY RISK FACTORS	QUANTIFY THE STEPS	PERFORM TASK/STEP BREAKDOWN	QUANTIFY EACH TASK	PERFORM DUTY/TASK BREAKDOWN	MEASURE DYNAMIC FORCES	MEASURE STATIC ELEMENTS	PERFORM ON-SITE OBSERVATION

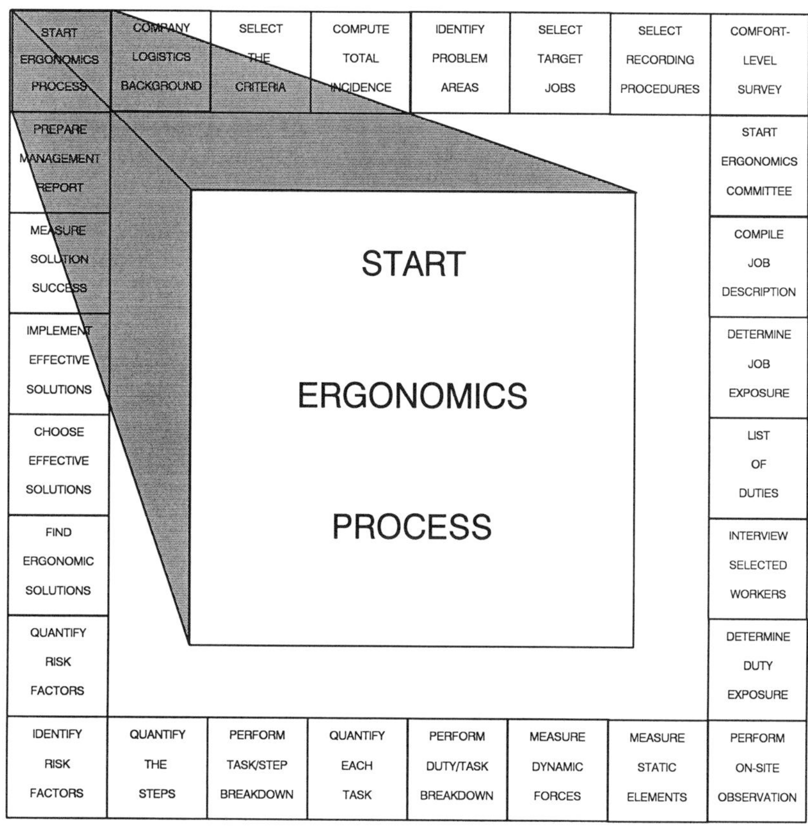
START ERGONOMICS PROCESS
COMPANY LOGISTICS BACKGROUND
SELECT THE CRITERIA
COMPUTE TOTAL INCIDENCE
IDENTIFY PROBLEM AREAS
SELECT TARGET JOBS
SELECT RECORDING PROCEDURES
COMFORT-LEVEL SURVEY
START ERGONOMICS COMMITTEE
COMPILE JOB DESCRIPTION
DETERMINE JOB EXPOSURE
LIST OF DUTIES
INTERVIEW SELECTED WORKERS
DETERMINE DUTY EXPOSURE
PERFORM ON-SITE OBSERVATION
MEASURE STATIC ELEMENTS
MEASURE DYNAMIC FORCES
PERFORM DUTY/TASK BREAKDOWN
QUANTIFY EACH TASK
PERFORM TASK/STEP BREAKDOWN
QUANTIFY THE STEPS
IDENTIFY RISK FACTORS
QUANTIFY RISK FACTORS
FIND ERGONOMIC SOLUTIONS
CHOOSE EFFECTIVE SOLUTIONS
IMPLEMENT EFFECTIVE SOLUTIONS
MEASURE SOLUTION SUCCESS
PREPARE MANAGEMENT REPORT
START
ERGONOMICS
PROCESS

Start Ergonomics Process

The overall goal of an ergonomic analysis is to provide information that will facilitate the development of an environment that enhances both human performance and well-being.

One objective is to generate a comprehensive report that will contain an assessment of specific workstations, identification of potentially hazardous conditions, and suggestions for improving the workstation. In most cases, that report will be presented to those decisionmakers in an organization with the authority to fund or direct necessary improvements in the workplace. In addition, those responsible persons will need to be kept informed of the progress, as well as the effectiveness, of the changes, improvements, or programs implemented as a part of this process.

In the introduction, seven steps were outlined to achieve this. Each of these steps is made up of smaller steps. The rest of this book will deal with these smaller steps. These steps can be arranged in relative order, similar to squares on a game board. As in the case of playing a board game, the best strategy is not always to land on every square. In some cases, you can achieve the same outcome in playing a game by only landing on some of the squares. It is not absolutely necessary to perform each one of the steps outlined here to meet the objective stated above. In a game, you try to move around the board in the proper order. In some cases, you have to back up and start again. In many cases, you can jump over several squares and progress faster without jeopardizing your chances of successfully completing the game.

As you move ahead with your ergonomics programs, you will also discover that certain steps can be "jumped over" without jeopardizing your success.

One of the most important aspects in the success of an ergonomics program is the creative nature of the ergonomic solutions. To facilitate that creativity, try to stay relaxed and have fun with the process. Think of it as a game. Each step has its own objective. Each step has its own procedures. Each step provides a new and exciting challenge. But all the steps lead to the one overall object of the game:

> To create an environment that enhances both human performance and well-being.

This is a simple approach. The procedures are based primarily on many years of practical experience. The basic principles can be traced back to numerous references. A bibliography can be found in the appendix.

The analogy between an ergonomics process and a board game can be expanded. To learn how to play a game, you can read all the instructions in the order that they are presented. Some people like to do that. Others like to just start playing and read the instructions as they land on each square. This book will facilitate either type of approach.

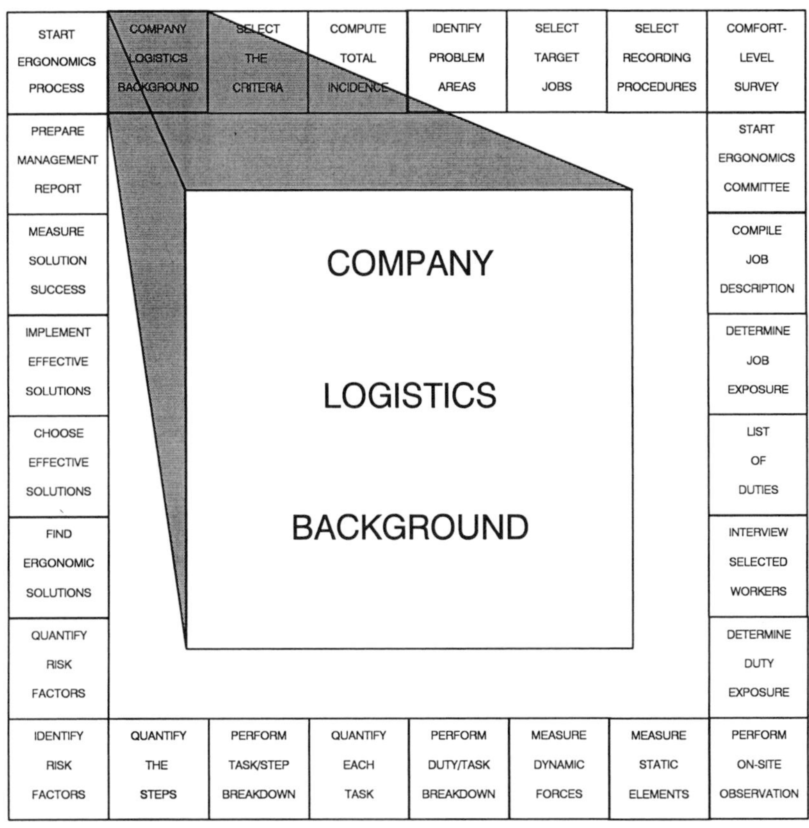
START ERGONOMICS PROCESS
COMPANY LOGISTICS BACKGROUND
SELECT THE CRITERIA
COMPUTE TOTAL INCIDENCE
IDENTIFY PROBLEM AREAS
SELECT TARGET JOBS
SELECT RECORDING PROCEDURES
COMFORT-LEVEL SURVEY
START ERGONOMICS COMMITTEE
COMPILE JOB DESCRIPTION
DETERMINE JOB EXPOSURE
LIST OF DUTIES
INTERVIEW SELECTED WORKERS
DETERMINE DUTY EXPOSURE
PERFORM ON-SITE OBSERVATION
MEASURE STATIC ELEMENTS
MEASURE DYNAMIC FORCES
PERFORM DUTY/TASK BREAKDOWN
QUANTIFY EACH TASK
PERFORM TASK/STEP BREAKDOWN
QUANTIFY THE STEPS
IDENTIFY RISK FACTORS
QUANTIFY RISK FACTORS
FIND ERGONOMIC SOLUTIONS
CHOOSE EFFECTIVE SOLUTIONS
IMPLEMENT EFFECTIVE SOLUTIONS
MEASURE SOLUTION SUCCESS
PREPARE MANAGEMENT REPORT
COMPANY LOGISTICS BACKGROUND

Company Logistics Background

OBJECTIVE

To get an accurate assessment of the current status of a company relative to its organization, policies, and injury-control efforts.

SUPPLIES

1. Company background forms
2. Agenda

GUIDELINES

This is the first chance you will have to foster or nurture management support. Without that support, the most comprehensive and concise ergonomics process will be a waste of time and effort. With that support, you will be able to develop a facility-wide team approach that will guarantee success. The overall goal is to get an accurate picture of the environment from the perspective of management.

This will allow you to rule out as many variables as possible regarding changes that either may have occurred in the past or will be taking place in the future. If there is a plan to change the current machinery or process, you will want to structure your report to assist in the redesign efforts.

PROCEDURES

Set up a meeting with representatives from as many of the following departments as possible:

Upper management
Facility management
Human resources
Health and safety
Engineering
Quality control
Production

The agenda may contain many of the components listed below. If at all possible, try to have current information regarding the cost of injuries and illness.

Generic Management Meeting Agenda

Definition of ergonomics
Current cost of injuries
Projected cost of injuries
Current and pending legislation regarding ergonomics
Action plan
Expected outcome
How it will affect them
Need for their support
Need for their input
Fill out company background form
Additional input regarding ergonomic hazards

Let them know what the goals and expected outcomes of the ergonomics program are. Share any past success you have had. Let them know what the mechanism for ongoing communication relative to your progress will be.

In an organized manner, ask questions relative to past and present policies, practices, the organization, and injury-control programs. The form that follows will facilitate this process.

It may be more productive to meet with representatives from all the departments individually, rather than as a group. Sometimes managers are less willing to be candid in a group setting. This may be due to concern that what they say will be held against them or that something they say may be interpreted as personal failure. Also it can be very difficult to coordinate everyone's schedules so that the appropriate amount of time is given to this topic.

This should be a meeting where the focus is clearly on the development of this process alone, rather than a 10-minute segment or a weekly 2-hour staff meeting.

COMPANY BACKGROUND QUESTIONNAIRE

1. Company Name *Ace Peanuts*

2. Type of Business *Peanut packers and distributors*

3. Number of Employees *75*

4. Number of Shifts *2*

5. Hours of Operation *6:00 a.m. to 12:00 midnight*

6. Days per Week of Operation *5 to 7*

7. Number of Hours per Shift *Varies*

8. Part Time Available *Yes* 9. Part Time Mandatory *Yes*

10. Overtime Available *No*

11. Overtime Mandatory *No*

12. Worker Compensated by (Hour?) Day? Quota? Salary? Other?

13. Rate Required? *No*

Significant Changes in Past 3 Years

Item Number	Significant Change
12 - Workers used to be compensated based on Pounds packed	

INITQ1

COMPANY BACKGROUND QUESTIONNAIRE

14. Number of Work Areas		Number of Workers
14a.	*Pickers*	*20*
14b.	*Material handlers*	*7*
14c.	*Inspectors*	*10*
14d.	*Bakers*	*8*
14e.	*Packers*	*6*
14f.	*Shipping*	*5*
14g.	*Clerical support*	*8*
14h.	*Management*	*10*
14i.		
14j.		
14k.		
14l.		

Significant Changes in Past 3 Years

Item Number	Significant Change
14a	*Number of pickers increased from 13 to 20*

INITQ2

THE BURKE PROTOCOL

COMPANY BACKGROUND QUESTIONNAIRE

15. Union Representation *No*

16. Transfers within Company Available *Yes*

17. Transfer Based on Seniority *Yes*

18. Transfer Mandatory *No*

19. Dress Codes / Uniforms *No*

20. Protective Equipment Required *No*

21. Training Program for New Employees *No*

22. Training Program for Transferred Employees *No*

23. Smoking Policy *No*

24. Muzak *No* 25. Radios allowed *Played Softly*

26. Eating / Drinking at Workstation Policy *No*

27. Cafeteria Available On Site *No*

Significant Changes in Past 3 Years

Item Number	Significant Change
	No changes

INITQ3

COMPANY BACKGROUND QUESTIONNAIRE

28. Incidence of Employee Complaints *Unknown*

29. Types of Injuries

Low back

Wrist

30. Costs of Injuries

Unknown

32. Injured Workers Cases Management *None*

33. Light Duty Available *No*

34. Early Return-to-Work Program Available *No*

35. Maximum Time on Light-Duty or Return-to-Work Program *N.A.*

36. Medical Facilities Available On Site *No*

Significant Changes in Past 3 Years

Item Number	Significant Change
29	*Recent increase in wrist pain complaints*

INITQ4

THE BURKE PROTOCOL

COMPANY BACKGROUND QUESTIONNAIRE

37. Changes in Equipment	*No*
38. Changes in Management	*No*
39. Changes in Safety Policy	*No*
40. Changes in Business Growth	*Yes, Business Growth Flat*
41. Changes in Facility Size	*No*
42. OSHA Inspections	*No*
43. Inspection Results	*N.A.*
44. Changes as a Result of O.S.H.A. Findings	—
45. Ergonomics Surveys Performed	*No*
46. Interventions Implemented	*No*
47. Success of Interventions	—
48. Injury Prevention Programs (Educational)	*No*
49. Injury Prevention Programs (Incentive)	*No*
50. Injury Prevention Programs (Other)	*No*

Item Number	Significant Changes

INITQ5

fsd

COMPANY BACKGROUND QUESTIONNAIRE

Item Number	Significant Changes
	No specific details necessary

INITQ6

THE BURKE PROTOCOL

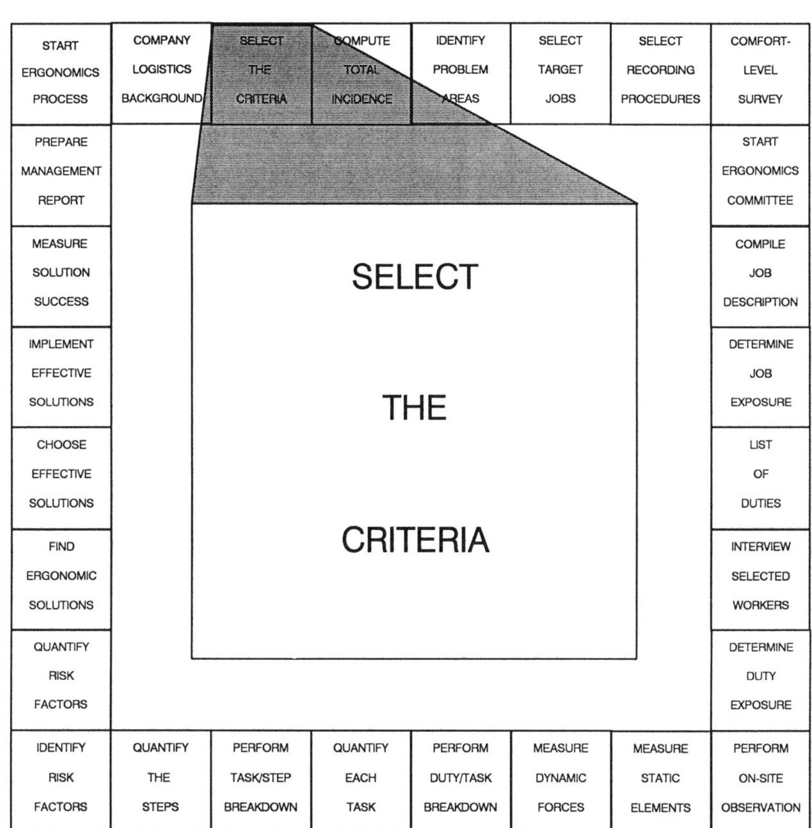

START ERGONOMICS PROCESS
COMPANY LOGISTICS BACKGROUND
SELECT THE CRITERIA
COMPUTE TOTAL INCIDENCE
IDENTIFY PROBLEM AREAS
SELECT TARGET JOBS
SELECT RECORDING PROCEDURES
COMFORT-LEVEL SURVEY
START ERGONOMICS COMMITTEE
COMPILE JOB DESCRIPTION
DETERMINE JOB EXPOSURE
LIST OF DUTIES
INTERVIEW SELECTED WORKERS
DETERMINE DUTY EXPOSURE
PERFORM ON-SITE OBSERVATION
MEASURE STATIC ELEMENTS
MEASURE DYNAMIC FORCES
PERFORM DUTY/TASK BREAKDOWN
QUANTIFY EACH TASK
PERFORM TASK/STEP BREAKDOWN
QUANTIFY THE STEPS
IDENTIFY RISK FACTORS
QUANTIFY RISK FACTORS
FIND ERGONOMIC SOLUTIONS
CHOOSE EFFECTIVE SOLUTIONS
IMPLEMENT EFFECTIVE SOLUTIONS
MEASURE SOLUTION SUCCESS
PREPARE MANAGEMENT REPORT
SELECT THE CRITERIA

Select the Criteria

OBJECTIVE

To determine the most effective and practical criteria for measuring the need for ergonomics study and intervention.

SUPPLIES

1. Any and all records available that reflect injury incidence, worker satisfaction, productivity, or quality control and that can act as a guide for choosing areas or jobs to analyze
2. Calculator

GUIDELINES

While a proactive approach to injury reduction is ideal, the driving force behind most programs is historical data. Often the issue of where to start looking for ergonomics conditions in need of attention is obvious. Your focus will be directed as a reaction to a known condition. Some jobs have a reputation for being extremely stressful. This subjective impression can usually be substantiated by a more formal record. This can be something as official as the government-mandated OSHA 200 record of illness and injuries form or something as informal as coffee-room gossip overheard by a line supervisor.

It is important to have an objective rationale for starting in a particular area. In addition, you will eventually be put in a position to make suggestions for reducing injuries. Following the implementation of those suggestions, you will want to track the effectiveness of that intervention. To facilitate this rationale, you need to establish a set of criteria.

The simplest method is to set specific criteria, utilize those criteria for prioritizing jobs, implement the changes to those jobs, track the effectiveness of the changes, and, hopefully, report that the changes in those criteria were favorable.

An example of this is utilizing the OSHA 200 record to break down the recordable injuries by department. Whichever department has the most injuries is where you should initiate your program. Following an analysis and the implementation of an intervention, you should look at the OSHA 200 logs again. Your efforts should be rewarded with a reduction of the number of injuries in that area.

This process is not always this simple. The criteria used for the selection of target jobs may not be sensitive enough to periodically assess the effectiveness of implemented interventions. For example, you may use the OSHA 200 log to determine in which area to initially concentrate your analyses. You may then come up with a new set of job-specific criteria, such as qualitative worker input, to assess the effectiveness of the interventions.

PROCEDURES

1. Look over the various criteria options available.
2. Choose one or a combination of options.
3. Go to the next step.

In the following subsections are found guidelines for selecting criteria.

Can It be Stated as an Objective Measurable Unit?

A valid criterion must be a concrete statement that can be expressed in a numerical value. This will allow you to first compare one area in a plant with another area in the same plant. It will then allow you to compare the effects of an intervention using a numerical rating system. It is important that, regardless of the criterion chosen, the units for comparison are the same.

Examples of different types of criteria are listed in the objective criteria chart.

Objective Criteria:

Injury incidence rate
Specific injury incidence rate
Direct cost of injuries
Subjective comfort-level survey
Absenteeism
Turnover
Tardiness
Productivity quota variance
Quality control variance
Number of workers on light duty
Non-work-related injury incidence
Lost-time days
Worker compensation costs

Incidence rates for injury or illness can be reduced to a standard Bureau of Labor Statistics statement of the number of injuries per 200,000 manhours.(See the next section for the formula.) This is fairly simple. The incidence rate is a statement of the ratio of the number of incidents per 100 employees working 40 hours per week for 50 weeks. In other words, it is a projection of the number of incidents if there were 100 workers working fifty 40-hour work weeks. An incidence rate of 2.0 would mean that 2 workers out of 100 were injured. It could also mean that 1 out of 50 or 20 out of 1000 workers were injured.

As mentioned earlier, worker attitudes can be a more sensitive assessment of ergonomic needs or progress. The challenge can be to take a relatively subjective impression of the worker and find a way to empirically track and record it. Attitude can be reflected in tardiness, absenteeism, or even visits to the medical department. A more direct assessment would be a job satisfaction survey or a worker comfort-level survey.

The final decision regarding criteria is to be made based on your unique set of circumstances. Remember, though, that if you are going to use worker input, you must get it before you initiate any interventions. This will establish a baseline for comparison after making a change. Examples of such data may be

> The subjective comfort level went from an average of 15 to an average of 20.
>
> Tardiness dropped from 15 to 3 incidents per 100 workers per week

This empirical approach is preferable to a purely subjective assessment, such as "Workers seem to be feeling better."

Is It Easy to Track?

The next consideration is extremely important and is often overlooked in the excitement to get a program underway. Think about who is going to be responsible for tracking this information. It is possible to devise a formula that attempts to take into account input from the health and safety department, human resources personnel, production management, and quality assurance personnel. While this formula would represent an excellent cross-section of information, it could be a full-time job tracking down all this information on a regular basis.

The OSHA 200 log is an easy-to-use and practical record. Many companies also keep records of visits to their medical department. This can also provide an accurate and practical criterion.

Before you get started, make sure you have determined who will be responsible for keeping track of the records and what records they will keep, to whom they will report, what format the report will take, and at what time intervals reports are to be generated.

In many cases the health and safety personnel will be the driving force behind implementation of many of the ergonomic interventions. Typically they already have the responsibility for maintenance of the OSHA 200 log and have access to other information regarding lost-time days and nonrecordable injuries. It is not unusual for the health and safety department personnel to act as administrators of an ergonomics program as part of their overall duties.

Can It be Converted to Dollars?

In many cases you will need to convert all the various considerations and criteria into a unit to which everyone can relate. This least common denominator is often money.

> Productivity people can say that productivity is up by 15 widgets per hour. What does that mean to human resource people?
>
> Human resource people can say that turnover has decreased by 23%. What does that mean to health and safety people?
>
> Health and safety people can say that injury rates are down by 36%. What does that mean to productivity people?

While all of the individuals named above understand that this is an improvement, the unit that translates across departments is financial impact. If you can state how much of the production costs for a product are the result of injuries, you can help to foster management commitment to an ergonomics intervention program.

In addition, it is a convenient way to compare the cost of injuries in one area to the cost in another area, and therefore, to help direct ergonomics intervention priorities. Most importantly, if you can reduce various criteria to a dollar amount and then demonstrate the change, you can help to rationalize the investment by stating it as a return on an investment.

In the cases stated above, this is good news. The adjective *good* is best seen by restating the results:

> The productivity statement represents an increase in profit of $3000.
>
> The human resource savings are equal to $7000.
>
> The injury rate reduction means a savings of $26,000.

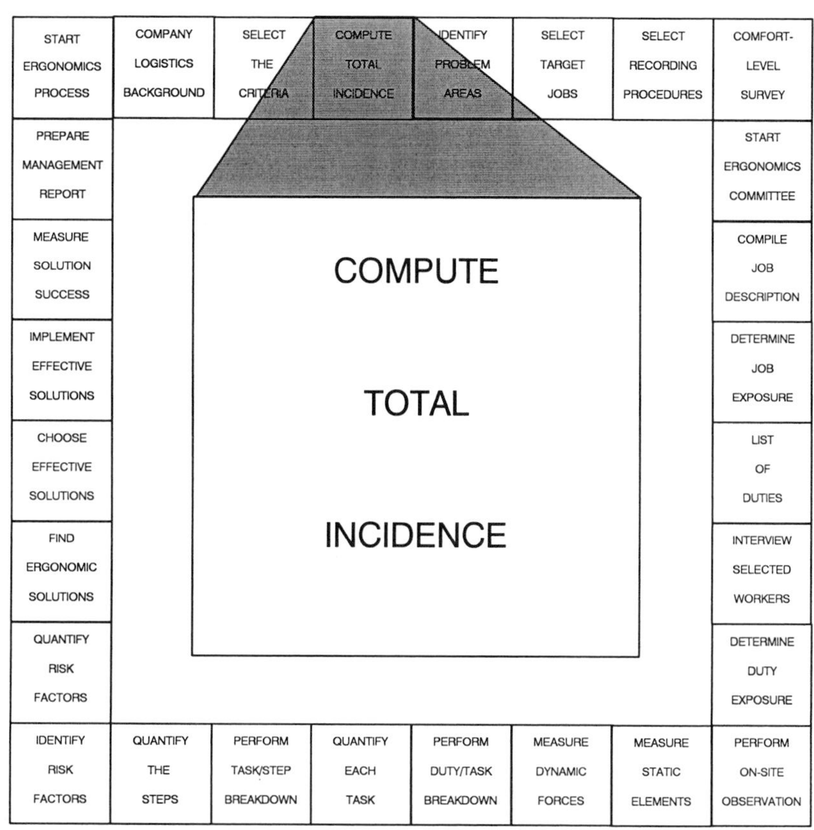
START ERGONOMICS PROCESS
COMPANY LOGISTICS BACKGROUND
SELECT THE CRITERIA
COMPUTE TOTAL INCIDENCE
IDENTIFY PROBLEM AREAS
SELECT TARGET JOBS
SELECT RECORDING PROCEDURES
COMFORT-LEVEL SURVEY
START ERGONOMICS COMMITTEE
COMPILE JOB DESCRIPTION
DETERMINE JOB EXPOSURE
LIST OF DUTIES
INTERVIEW SELECTED WORKERS
DETERMINE DUTY EXPOSURE
PERFORM ON-SITE OBSERVATION
MEASURE STATIC ELEMENTS
MEASURE DYNAMIC FORCES
PERFORM DUTY/TASK BREAKDOWN
QUANTIFY EACH TASK
PERFORM TASK/STEP BREAKDOWN
QUANTIFY THE STEPS
IDENTIFY RISK FACTORS
QUANTIFY RISK FACTORS
FIND ERGONOMIC SOLUTIONS
CHOOSE EFFECTIVE SOLUTIONS
IMPLEMENT EFFECTIVE SOLUTIONS
MEASURE SOLUTION SUCCESS
PREPARE MANAGEMENT REPORT
COMPUTE
TOTAL
INCIDENCE

Compute Total Incidence

OBJECTIVE

To arrive at the injury incidence rate for the entire facility.

SUPPLIES

1. OSHA 200 log or other injury and illness records
2. Information on
   - average number of employees
   - average number of hours worked per week
   - average number of weeks worked per year
3. Form "How to Compute Incidence Rate"
4. Calculator

GUIDELINES

The OSHA 200 log is a federally mandated recordkeeping form. Rules vary as to which injury or illness incidence must be recorded. Severe penalties result from failure to comply with this procedure. As a result, it can be a simple and accurate device for calculating the incidence rate.

Incidence rates for injury or illness can be reduced to a standard Bureau of Labor Statistics statement of the number of injuries per 200,000 manhours.

Procedures

Calculate total number of hours worked in a year:

1. Write down the total number of employees.
2. Write down the number of hours each employee worked per week.
3. Write down the number of weeks worked in a year.
4. Multiply the number in steps 1–3.

This is the total number of hours worked in a year.

5. Write down the total number of injuries (or events) that occurred in a year.
6. Multiply the total number of injuries in step 5 by 200,000.
7. Divide the number from step 6 by the number from step 4.

How to Compute Incidence Rate

$$\text{Incidence Rate} = \frac{(1)\ \text{Number of injuries of illnesses} \times 200{,}000}{(2)\ \text{Actual total number of hours worked by all employees}}$$

1. The number of illnesses or injuries can be arrived at by counting the number of entries on the OSHA 200 form or by adding columns 1, 2, 6, 8, 9, and 13.
2. The total number of hours worked can be found in payroll accounting records or by multiplying number of employees by number of hours worked per day by number of days worked per year.

Shortcut

If your company averages between 48 and 52 weeks worked per year and most positions are worked 40 hours a week, you might want to use the tables at the end of this chapter to do a quick conversion. If you are already experienced and comfortable with the process of calculating the incidence rate, you will not need them.

Locate the index number in the tables. This number is used to multiply the current number of injuries to arrive at the incidence rate. These tables are used as ergonomic criteria and not as a substitution for OSHA recording.

Do Not Use These Tables For Any Official Reporting of The OSHA Incidence Rate. These Tables Are An Approximation.

This is very important and cannot be overstated.

Do Not Use These Tables For Any Official Reporting of The OSHA Incidence Rate. These Tables Are An Approximation.

In most cases you will not know the exact number of employees working in a year. There is almost always some shifting around of employees. As with many other procedures, you can try to determine an average number of positions worked in a year. In other words, if you have 10 positions available, but in a single year 3 people left and were replaced by 3 other people, you would still only consider the 10 positions and not the 13 workers.

Often the human resources department can give you the actual number of hours worked in a given year.

You can also use this format with other objective records. For example, you might try looking at reject levels as an objective measurement of worker comfort. You can compute the incidence of rejects per 100 workers working 200,000 manhours before and after you initiate an ergonomics intervention.

Injury Incidence Index Numbers

1. Choose the number of workers in the right column.
2. Find the index number based on the number of weeks/year worked.
3. Multiply the index number by the number of injuries.

Total # Injuries X Index Number = Incidence rate

Number of Workers	52 Weeks/Year	51 Weeks/Year	50 Weeks/Year	49 Weeks/Year	48 Weeks/Year
10	9.62	9.80	10.00	10.20	10.42
20	4.81	4.90	5.00	5.10	5.21
30	3.21	3.27	3.33	3.40	3.47
40	2.40	2.45	2.50	2.55	2.60
50	1.92	1.96	2.00	2.04	2.08
60	1.60	1.63	1.67	1.70	1.74
70	1.37	1.40	1.43	1.46	1.49
80	1.20	1.23	1.25	1.28	1.30
90	1.07	1.09	1.11	1.13	1.16
100	0.96	0.98	1.00	1.02	1.04
110	0.87	0.89	0.91	0.93	0.95
120	0.80	0.82	0.83	0.85	0.87
130	0.74	0.75	0.77	0.78	0.80
140	0.69	0.70	0.71	0.73	0.74
150	0.64	0.65	0.67	0.68	0.69
160	0.60	0.61	0.63	0.64	0.65
170	0.57	0.58	0.59	0.60	0.61
180	0.53	0.54	0.56	0.57	0.58
190	0.51	0.52	0.53	0.54	0.55
200	0.48	0.49	0.50	0.51	0.52
210	0.46	0.47	0.48	0.49	0.50
220	0.44	0.45	0.45	0.46	0.47
230	0.42	0.43	0.43	0.44	0.45
240	0.40	0.41	0.42	0.43	0.43
250	0.38	0.39	0.40	0.41	0.42
260	0.37	0.38	0.38	0.39	0.40
270	0.36	0.36	0.37	0.38	0.39
280	0.34	0.35	0.36	0.36	0.37
290	0.33	0.34	0.34	0.35	0.36
300	0.32	0.33	0.33	0.34	0.35
310	0.31	0.32	0.32	0.33	0.34
320	0.30	0.31	0.31	0.32	0.33
330	0.29	0.30	0.30	0.31	0.32
340	0.28	0.29	0.29	0.30	0.31
350	0.27	0.28	0.29	0.29	0.30
360	0.27	0.27	0.28	0.28	0.29

(Based on 40 hour work week)

Injury Incidence Index Numbers

1. Choose the number of workers in the right column.
2. Find the index number based on the number of weeks/year worked.
3. Multiply the index number by the number of injuries.

Total # Injuries X Index Number = Incidence rate

Number of Workers	52 Weeks/Year	51 Weeks/Year	50 Weeks/Year	49 Weeks/Year	48 Weeks/Year
370	0.26	0.26	0.27	0.28	0.28
380	0.25	0.26	0.26	0.27	0.27
390	0.25	0.25	0.26	0.26	0.27
400	0.24	0.25	0.25	0.26	0.26
410	0.23	0.24	0.24	0.25	0.25
420	0.23	0.23	0.24	0.24	0.25
430	0.22	0.23	0.23	0.24	0.24
440	0.22	0.22	0.23	0.23	0.24
450	0.21	0.22	0.22	0.23	0.23
460	0.21	0.21	0.22	0.22	0.23
470	0.20	0.21	0.21	0.22	0.22
480	0.20	0.20	0.21	0.21	0.22
490	0.20	0.20	0.20	0.21	0.21
500	0.19	0.20	0.20	0.20	0.21
510	0.19	0.19	0.20	0.20	0.20
520	0.18	0.19	0.19	0.20	0.20
530	0.18	0.18	0.19	0.19	0.20
540	0.18	0.18	0.19	0.19	0.19
550	0.17	0.18	0.18	0.19	0.19
560	0.17	0.18	0.18	0.18	0.19
570	0.17	0.17	0.18	0.18	0.18
580	0.17	0.17	0.17	0.18	0.18
590	0.16	0.17	0.17	0.17	0.18
600	0.16	0.16	0.17	0.17	0.17
610	0.16	0.16	0.16	0.17	0.17
620	0.16	0.16	0.16	0.16	0.17
630	0.15	0.16	0.16	0.16	0.17
640	0.15	0.15	0.16	0.16	0.16
650	0.15	0.15	0.15	0.16	0.16
660	0.15	0.15	0.15	0.15	0.16
670	0.14	0.15	0.15	0.15	0.16
680	0.14	0.14	0.15	0.15	0.15
690	0.14	0.14	0.14	0.15	0.15
700	0.14	0.14	0.14	0.15	0.15
710	0.14	0.14	0.14	0.14	0.15
720	0.13	0.14	0.14	0.14	0.14

(Based on 40 hour work week)

Injury Incidence Index Numbers

1. Choose the number of workers in the right column.
2. Find the index number based on the number of weeks/year worked.
3. Multiply the index number by the number of injuries.

Total # Injuries X Index Number = Incidence rate

Number of Workers	52 Weeks/Year	51 Weeks/Year	50 Weeks/Year	49 Weeks/Year	48 Weeks/Year
730	0.13	0.13	0.14	0.14	0.14
740	0.13	0.13	0.14	0.14	0.14
750	0.13	0.13	0.13	0.14	0.14
760	0.13	0.13	0.13	0.13	0.14
770	0.12	0.13	0.13	0.13	0.14
780	0.12	0.13	0.13	0.13	0.13
790	0.12	0.12	0.13	0.13	0.13
800	0.12	0.12	0.13	0.13	0.13
810	0.12	0.12	0.12	0.13	0.13
820	0.12	0.12	0.12	0.12	0.13
830	0.12	0.12	0.12	0.12	0.13
840	0.11	0.12	0.12	0.12	0.12
850	0.11	0.12	0.12	0.12	0.12
860	0.11	0.11	0.12	0.12	0.12
870	0.11	0.11	0.11	0.12	0.12
880	0.11	0.11	0.11	0.12	0.12
890	0.11	0.11	0.11	0.11	0.12
900	0.11	0.11	0.11	0.11	0.12
910	0.11	0.11	0.11	0.11	0.11
920	0.10	0.11	0.11	0.11	0.11
930	0.10	0.11	0.11	0.11	0.11
940	0.10	0.10	0.11	0.11	0.11
950	0.10	0.10	0.11	0.11	0.11
960	0.10	0.10	0.10	0.11	0.11
970	0.10	0.10	0.10	0.11	0.11
980	0.10	0.10	0.10	0.10	0.11
990	0.10	0.10	0.10	0.10	0.11
1000	0.10	0.10	0.10	0.10	0.10

(Based on 40 hour work week)

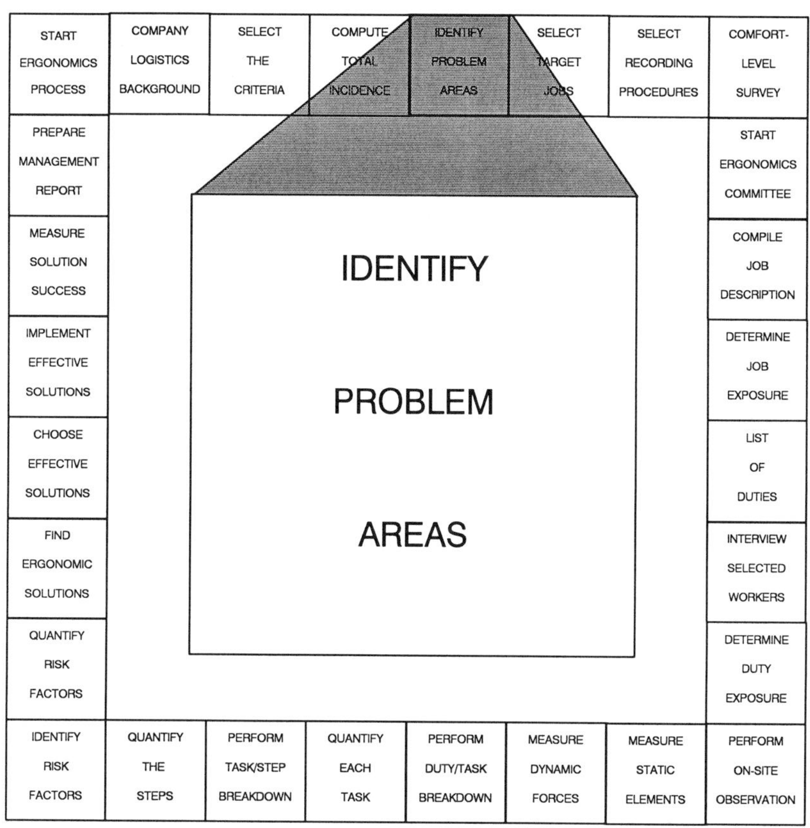

START ERGONOMICS PROCESS
COMPANY LOGISTICS BACKGROUND
SELECT THE CRITERIA
COMPUTE TOTAL INCIDENCE
IDENTIFY PROBLEM AREAS
SELECT TARGET JOBS
SELECT RECORDING PROCEDURES
COMFORT-LEVEL SURVEY
START ERGONOMICS COMMITTEE
COMPILE JOB DESCRIPTION
DETERMINE JOB EXPOSURE
LIST OF DUTIES
INTERVIEW SELECTED WORKERS
DETERMINE DUTY EXPOSURE
PERFORM ON-SITE OBSERVATION
MEASURE STATIC ELEMENTS
MEASURE DYNAMIC FORCES
PERFORM DUTY/TASK BREAKDOWN
QUANTIFY EACH TASK
PERFORM TASK/STEP BREAKDOWN
QUANTIFY THE STEPS
IDENTIFY RISK FACTORS
QUANTIFY RISK FACTORS
FIND ERGONOMIC SOLUTIONS
CHOOSE EFFECTIVE SOLUTIONS
IMPLEMENT EFFECTIVE SOLUTIONS
MEASURE SOLUTION SUCCESS
PREPARE MANAGEMENT REPORT
IDENTIFY
PROBLEM
AREAS

Identify Problem Areas

OBJECTIVE

To utilize existing records to determine the greatest need for an ergonomic intervention.

SUPPLIES

1. Calculator
2. OSHA 200 log
3. Other injury and illness records

GUIDELINES

Up to this point you have been looking at the record and performance of your facility as a whole unit. This has provided you with insight to determine if there is a historical basis for establishing the need for initiating an ergonomics process in your facility. The next step is to determine where in the facility the greatest need exists.

PROCEDURES

1. List by department, area, or job description the number of injuries, as well as the incidence rate, for each area.
2. Calculate the ratio of the number of injuries per worker.

To determine the incidence rate for each department or job description, you will only need to calculate it as a ratio of the number of injuries to the number of employees in the department.

$$\text{Injury ratio} = \frac{\text{Number of injuries}}{\text{Number of workers in the department}} \times 100$$

Example

Ace Company Inc.
Record of Injuries (1987)

Area	# Employees	# Injuries	Rate
Receiving	9	—	—
Pickers	21	6	28.6
Product prep	25	1	4.0
Bakers	33	4	12.1
Packers	21	—	—
Inspectors	11	—	—
Office support	20	1	5.0
Sales	16	—	—
Total	156	12	7.68

*Incidence rate based on 40 hours per week and 50 weeks per year worked.

3. Arrange the various departments, areas, or job titles in order of greatest incidence rate.

Example

Job Titles Arranged by Number of Injuries

Area	# Employees	# Injuries	Rate
Pickers	21	6	28.6
Bakers	33	4	12.1
Product prep	25	1	4.0
Office support	20	1	5.0

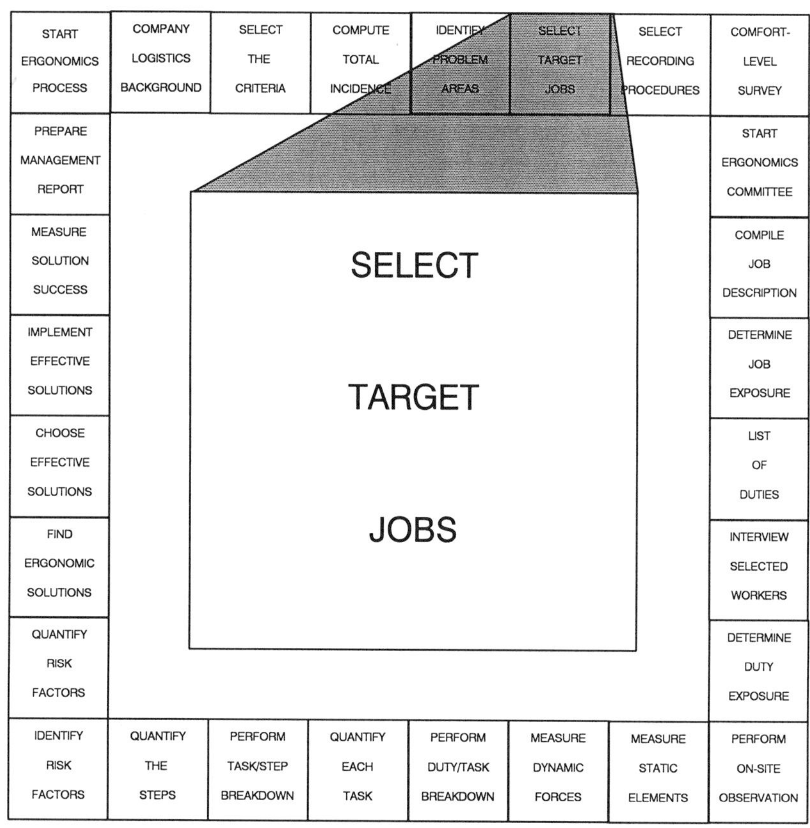
START ERGONOMICS PROCESS
COMPANY LOGISTICS BACKGROUND
SELECT THE CRITERIA
COMPUTE TOTAL INCIDENCE
IDENTIFY PROBLEM AREAS
SELECT TARGET JOBS
SELECT RECORDING PROCEDURES
COMFORT-LEVEL SURVEY
START ERGONOMICS COMMITTEE
COMPILE JOB DESCRIPTION
DETERMINE JOB EXPOSURE
LIST OF DUTIES
INTERVIEW SELECTED WORKERS
DETERMINE DUTY EXPOSURE
PERFORM ON-SITE OBSERVATION
MEASURE STATIC ELEMENTS
MEASURE DYNAMIC FORCES
PERFORM DUTY/TASK BREAKDOWN
QUANTIFY EACH TASK
PERFORM TASK/STEP BREAKDOWN
QUANTIFY THE STEPS
IDENTIFY RISK FACTORS
QUANTIFY RISK FACTORS
FIND ERGONOMIC SOLUTIONS
CHOOSE EFFECTIVE SOLUTIONS
IMPLEMENT EFFECTIVE SOLUTIONS
MEASURE SOLUTION SUCCESS
PREPARE MANAGEMENT REPORT
SELECT TARGET JOBS

Select Target Jobs

OBJECTIVE

To develop a schedule for the analysis, report, and implementation of ergonomic process in each area.

SUPPLIES

1. A breakdown of the most hazardous areas or job descriptions based on history of injuries or other chosen criteria.

GUIDELINES

It is important to have an objective rationale to explain your decision regarding in which department, area, or workstation you want to start the ergonomics process. This objective rationale should then be used in a consistent manner to prioritize and direct your projects. It is most effective to equate the objective rationale to a cost factor. For example, show that the amount of money paid out in worker compensation for the previous year is equal to the profits for the first 2 months of this year.

In addition, it is helpful, at this point, to develop a schedule of when each area will be analyzed, when a report will be presented based on that analysis, and when suggestions from that report will be implemented. This can help to prevent managers or departments from feeling that they are being singled out or picked on. Each facility and company will vary greatly in terms of scheduling. Try to be as aggressive as possible while still remaining realistic.

PROCEDURES

1. Gather all the objective information you have that compares all the departments in the facility.
2. Choose the appropriate criteria.
3. Compile a schedule of job titles to be analyzed.

HINTS

The selection of the initial areas in which to implement the ergonomics process can be based on a variety of factors. The first and simplest is the record of injuries. A second and extremely important criterion is your perception of the chances for success in an area. It is important, especially at first, to successfully implement the process and to get the desired result. This result may be increased productivity, fewer injuries, higher and more consistent quality, or better morale. You need to consider the level of commitment of the management in that area. If you are going to have a hard time getting time to talk to the workers or videotaping the workers, then you can anticipate that when it is time to implement an intervention the cooperation will be minimal. Other factors that may come into play are quality, productivity, or labor-relations issues.

The following example shows how simple the schedule can be. It calls for a different job to be analyzed each month. A 1-month period is provided to prepare the report. This report is presented to the decisionmakers in an organization. Hopefully a decision can be made and implementation facilitated within 2 months after the report is given. This would be the case for a highly motivated management team.

Example

Schedule of Job Titles and Areas for Ergonomics Analysis (1988)

Based on the records of injury and illness incidence, the following areas have been scheduled for ergonomic analysis:

Area	Analysis	Report	Implementation
Pickers	June 1	July 1	Sept 1
Bakers	July 1	Aug 1	Oct 1
Product prep	Aug 1	Sept 1	Jan 1 '89
Office support	Sept 1	Oct 1	Feb 1 '89

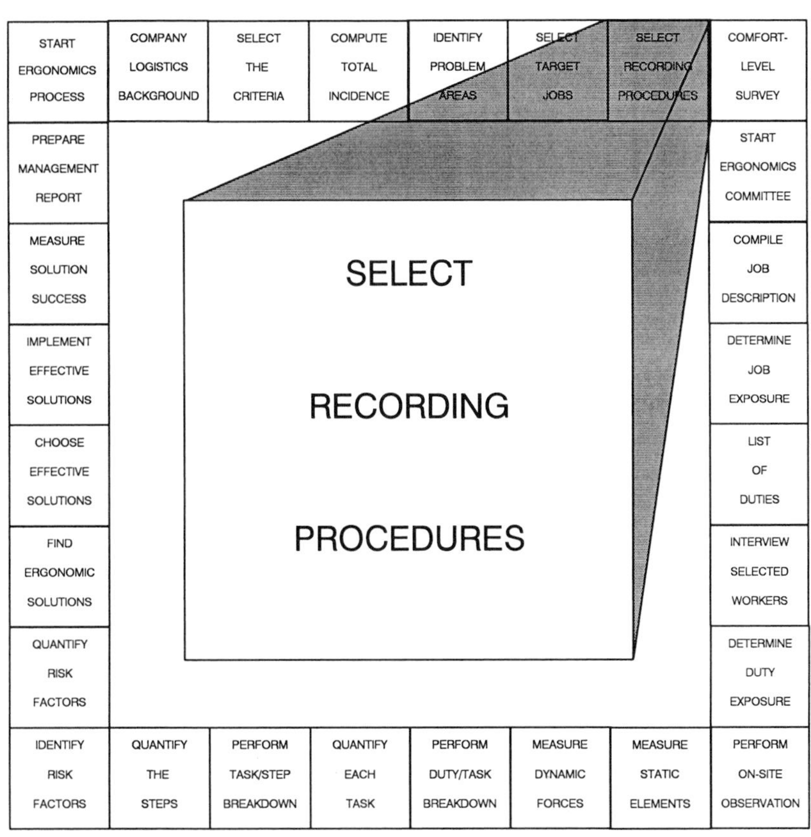
START ERGONOMICS PROCESS
COMPANY LOGISTICS BACKGROUND
SELECT THE CRITERIA
COMPUTE TOTAL INCIDENCE
IDENTIFY PROBLEM AREAS
SELECT TARGET JOBS
SELECT RECORDING PROCEDURES
COMFORT-LEVEL SURVEY
START ERGONOMICS COMMITTEE
COMPILE JOB DESCRIPTION
DETERMINE JOB EXPOSURE
LIST OF DUTIES
INTERVIEW SELECTED WORKERS
DETERMINE DUTY EXPOSURE
PERFORM ON-SITE OBSERVATION
MEASURE STATIC ELEMENTS
MEASURE DYNAMIC FORCES
PERFORM DUTY/TASK BREAKDOWN
QUANTIFY EACH TASK
PERFORM TASK/STEP BREAKDOWN
QUANTIFY THE STEPS
IDENTIFY RISK FACTORS
QUANTIFY RISK FACTORS
FIND ERGONOMIC SOLUTIONS
CHOOSE EFFECTIVE SOLUTIONS
IMPLEMENT EFFECTIVE SOLUTIONS
MEASURE SOLUTION SUCCESS
PREPARE MANAGEMENT REPORT
SELECT RECORDING PROCEDURES

Select Recording Procedures

OBJECTIVE

To determine and document the recording procedures and responsibilities associated with the ergonomics process.

SUPPLIES

1. Criteria Worksheet

GUIDELINES

It is vital to determine who will be responsible for keeping track of the records. Try to be as comprehensive in this step as possible. You do not want to get a year down the road after having implemented numerous ergonomic interventions and then have to backtrack to measure the effectiveness.

The types of decisions that need to be made are

What records are to be kept?

To whom will the report be sent?

What format will the report take?

At what time intervals will the reports be generated?

PROCEDURES

1. Have a meeting with the person or persons who will be responsible for the recordkeeping.
2. Jointly discuss realistic expectations for recordkeeping and reporting procedures.
3. Document the results of that meeting.

Communicate frequently with the individuals responsible for the recordkeeping to provide support, guidance, and motivation. Not everyone understands the importance of these types of records.

The form that follows can be helpful in documenting everyone's responsibilities.

Page 1

CRITERIA WORKSHEET

1. WHO WILL BE RESPONSIBLE FOR COMPILING INFORMATION ?

2. WHAT ARE THE CRITERIA TO BE USED FOR CHOOSING THE TARGET JOB POSITION?

- X OSHA RECORDABLE INCIDENCE
- ___ PRODUCTIVITY VARIANCES
- ___ TURNOVER RATE
- ___ ALL INJURIES (ON DUTY) EVEN NON- RECORDABLE
- ___ NUMBER OF TOTAL INJURIES
- ___ TOTAL MEDICAL COSTS OF INJURIES
- ___ LOST WORK DAYS
- ___ LOST-TIME INJURIES ONLY
- ___ TOTAL ESTIMATED COST BASED ON FORMULA TO BE DEVELOPED
- ___ NUMBER OF VISITS TO MEDICAL DEPARTMENT
- ___ REJECT VARIANCE REPORTS
- X SUBJECTIVE COMFORT-LEVEL SURVEY RESULTS

3. HOW OFTEN WILL INFORMATION BE COMPILED?

- ___ MONTHLY
- ___ EVERY 2 MONTHS
- X QUARTERLY
- ___ EVERY 6 MONTHS
- ___ YEARLY

THE BURKE PROTOCOL F-CRITR1

Page 2

CRITERIA WORKSHEET

5. WHO WILL BE RESPONSIBLE FOR SENDING THIS TO THE LOCAL PERSON?

Reporting Information	Source (Person's Name)Title
OSHA 200 Report	*Human resources*
Total hours worked	*Human resources*

6. WHAT IS THE FORMAT OF THE REPORT TO BE GENERATED?

_X_ COMPARISON OF YEAR TO DATE

____ COMPARISON OF PAST SEVERAL YEARS

____ CHANGE SINCE PROJECT INITIATION

____ COMPARISON TO SIMILAR POSITION

7. HOW WILL THE REPORT BE PRESENTED? _X_ WRITTEN ____ ORAL

8. TO WHOM WILL THIS INFORMATION BE SENT?

____ ERGONOMICS TASK FORCE

_X_ FACILITY MANAGER

_X_ COMPANY HEALTH AND SAFETY PERSONS

____ OTHER____________________

____ OTHER____________________

THE BURKE PROTOCOL FORM F-CRITR2

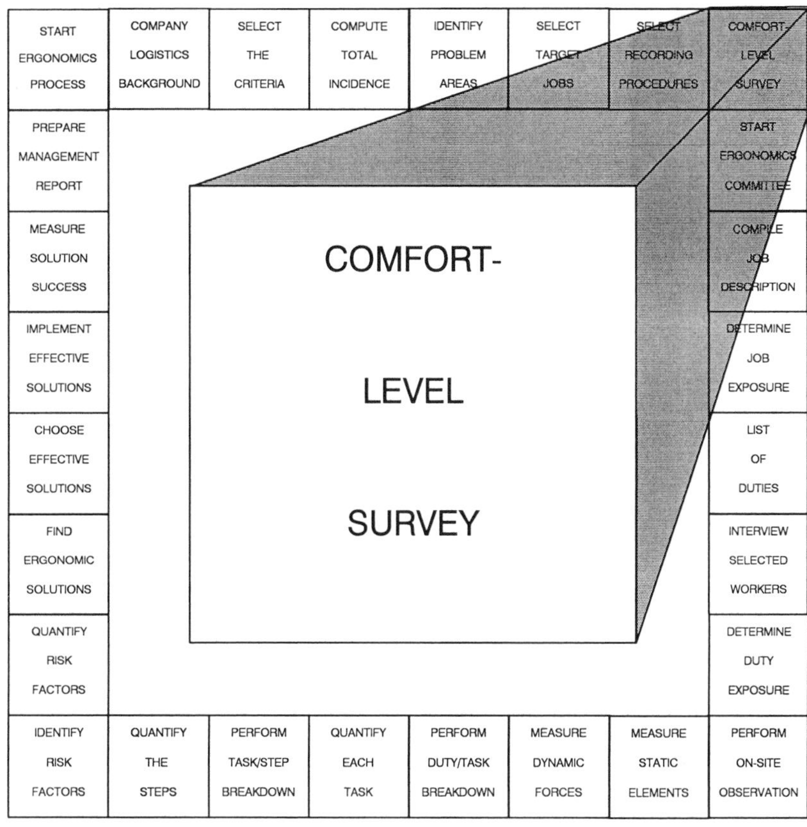
START
ERGONOMICS
PROCESS
COMPANY
LOGISTICS
BACKGROUND
SELECT
THE
CRITERIA
COMPUTE
TOTAL
INCIDENCE
IDENTIFY
PROBLEM
AREAS
SELECT
TARGET
JOBS
SELECT
RECORDING
PROCEDURES
COMFORT-
LEVEL
SURVEY
START
ERGONOMICS
COMMITTEE
COMPILE
JOB
DESCRIPTION
DETERMINE
JOB
EXPOSURE
LIST
OF
DUTIES
INTERVIEW
SELECTED
WORKERS
DETERMINE
DUTY
EXPOSURE
PERFORM
ON-SITE
OBSERVATION
MEASURE
STATIC
ELEMENTS
MEASURE
DYNAMIC
FORCES
PERFORM
DUTY/TASK
BREAKDOWN
QUANTIFY
EACH
TASK
PERFORM
TASK/STEP
BREAKDOWN
QUANTIFY
THE
STEPS
IDENTIFY
RISK
FACTORS
QUANTIFY
RISK
FACTORS
FIND
ERGONOMIC
SOLUTIONS
CHOOSE
EFFECTIVE
SOLUTIONS
IMPLEMENT
EFFECTIVE
SOLUTIONS
MEASURE
SOLUTION
SUCCESS
PREPARE
MANAGEMENT
REPORT
COMFORT-
LEVEL
SURVEY

Comfort-Level Survey

OBJECTIVE

To gather and analyze the subjective comfort level in a particular job title or area, and convert it to a measurable objective unit.

SUPPLIES

1. Comfort-level survey
2. Calculator

GUIDELINES

The purpose of performing this step is to get an objective measurement of worker subjective comfort. This can provide additional support for initiating the process in a specific area. It can also be used as a more sensitive measurement to assess the effectiveness of an intervention.

PROCEDURES

This can be a part of a regular employee meeting or a special event. It is important to let everyone know that the results are anonymous and that this is part of an overall effort to improve their jobs.

The instructions would be as follows:

> Please complete the following survey by circling the answer that best describes your experience. This form will be used to measure the effectiveness of our efforts to increase the comfort and safety of your job. This information is completely anonymous. Please be totally honest in your assessment.

One word of caution here: Unless you have access to some type of computerized data-collection equipment, you will have to perform the data compilation by hand. For that reason, try to keep information as simple as possible. The sample provided covers many more body parts than you are likely to test. You should limit the number to three to four body parts.

Here is an example of one way in which a subjective comfort level can be used. Just before initiating the analysis process in an office area, collect data relative to the comfort level.

Each response is given a value from 1 to 4, representing the frequency that the workers experience either mild or strong discomfort.

1. Rarely or never
2. One or two days per month
3. Several days per month
4. Every day or almost every day

Add up all the values for the responses in each question box and divide the sum by the total number of respondents. This will give you the average frequency that individuals experience discomfort in the body part in question.

The results are calculated for each body part and for either mild or strong discomfort. The final tally shows that workers are experiencing mild neck pain at a frequency level of 3.01 on several days per month and strong neck pain at a frequency level of 1.75, or somewhere between rarely and a couple of days per month.

COMFORT-LEVEL SURVEY

Date
Job title or number
Shift
I have worked in this job for

I experience MILD DISCOMFORT in my feet	I experience STRONG DISCOMFORT in my feet
☐ Rarely or never ☐ One or two days per month ☐ Several days per month ☐ Every day	☐ Rarely or never ☐ One or two days per month ☐ Several days per month ☐ Every day
I experience MILD DISCOMFORT in my knees	**I experience STRONG DISCOMFORT in my knees**
☐ Rarely or never ☐ One or two days per month ☐ Several days per month ☐ Every day	☐ Rarely or never ☐ One or two days per month ☐ Several days per month ☐ Every day
I experience MILD DISCOMFORT in my back	**I experience STRONG DISCOMFORT in my back**
☐ Rarely or never ☐ One or two days per month ☐ Several days per month ☐ Every day	☐ Rarely or never ☐ One or two days per month ☐ Several days per month ☐ Every day
I experience MILD DISCOMFORT in my neck	**I experience STRONG DISCOMFORT in my neck**
☐ Rarely or never ☐ One or two days per month ☐ Several days per month ☐ Every day	☐ Rarely or never ☐ One or two days per month ☐ Several days per month ☐ Every day
I experience MILD DISCOMFORT in my arms	**I experience STRONG DISCOMFORT in my arms**
☐ Rarely or never ☐ One or two days per month ☐ Several days per month ☐ Every day	☐ Rarely or never ☐ One or two days per month ☐ Several days per month ☐ Every day

SUBJECTIVE COMFORT-LEVEL SURVEY RESULTS: OFFICE AREA
June 5, 1972

Body Part	Mild	Strong
Feet	1.76	1.10
Knees	1.05	1.51
Back	3.11	1.99
Neck	3.01	1.75
Arms	2.55	1.09

After the analysis, it is decided to initiate rest breaks at all the computer terminals. About 3 months later, the test is repeated.

SUBJECTIVE COMFORT-LEVEL SURVEY RESULTS: OFFICE AREA
September 14, 1972

Body Part	Mild	Strong
Feet	1.07	1.05
Knees	1.15	1.01
Back	3.15	1.79
Neck	1.67	1.08
Arms	2.55	1.09

The results indicate that the comfort level of the neck has improved dramatically. It would have taken considerably longer to assess the effectiveness of this intervention if it had been based solely on injury and illness data.

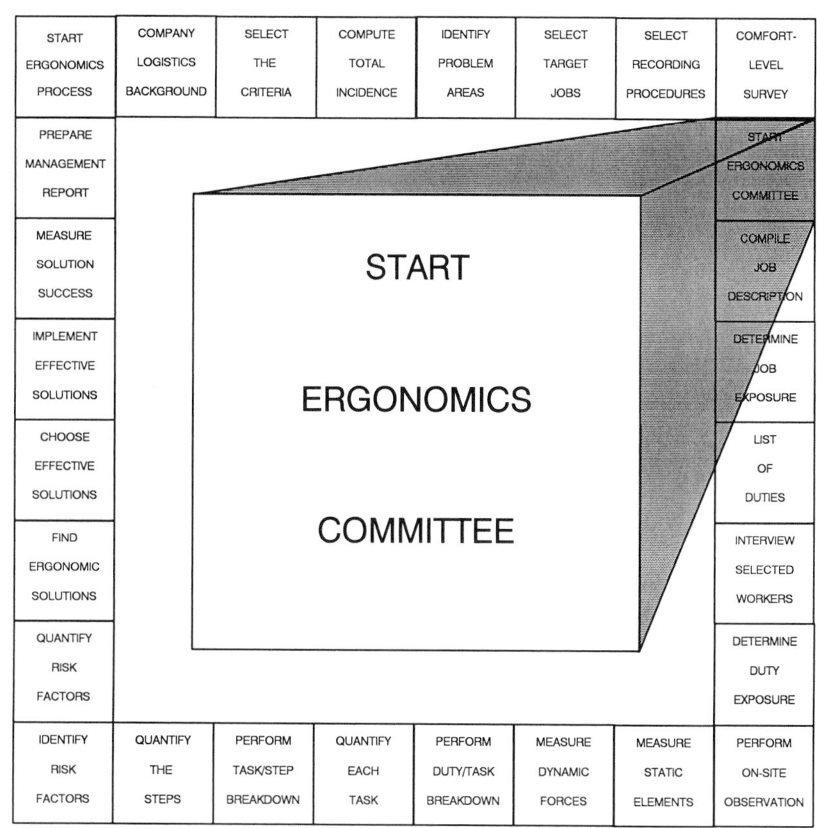

START ERGONOMICS PROCESS
COMPANY LOGISTICS BACKGROUND
SELECT THE CRITERIA
COMPUTE TOTAL INCIDENCE
IDENTIFY PROBLEM AREAS
SELECT TARGET JOBS
SELECT RECORDING PROCEDURES
COMFORT-LEVEL SURVEY
START ERGONOMICS COMMITTEE
COMPILE JOB DESCRIPTION
DETERMINE JOB EXPOSURE
LIST OF DUTIES
INTERVIEW SELECTED WORKERS
DETERMINE DUTY EXPOSURE
PERFORM ON-SITE OBSERVATION
MEASURE STATIC ELEMENTS
MEASURE DYNAMIC FORCES
PERFORM DUTY/TASK BREAKDOWN
QUANTIFY EACH TASK
PERFORM TASK/STEP BREAKDOWN
QUANTIFY THE STEPS
IDENTIFY RISK FACTORS
QUANTIFY RISK FACTORS
FIND ERGONOMIC SOLUTIONS
CHOOSE EFFECTIVE SOLUTIONS
IMPLEMENT EFFECTIVE SOLUTIONS
MEASURE SOLUTION SUCCESS
PREPARE MANAGEMENT REPORT
START
ERGONOMICS
COMMITTEE

Start Ergonomics Committee

OBJECTIVE

To set up a local ergonomics committee.

SUPPLIES

None

GUIDELINES

The purpose of utilizing a committee is to get the maximum input from the various persons who will be affected by any changes to the workstation. In addition, it provides some credibility to any suggestions. Most workers have an intuitive expertise in their jobs. They have the most to lose by ignoring harmful conditions, as well as the most to gain from improving their working conditions.

The ergonomics committee will work best if kept to about eight members. It is critical that the individuals chosen to participate in the committee understand the importance of their commitment. This may require additional uncompensated time on their part. It should be firmly understood that this is not a way to get out of work.

Ideally the committee should be made up of either the same number or slightly more hourly workers as non-hourly workers. The ideal profile is individuals who are imaginative, dedicated, and hard working. Finding such individuals is truly the challenge of this step. Often individuals with a history of a musculoskeletal injury make good participants. A healthy mix of typically vocal and typically quiet workers helps to keep things from getting too out of hand.

It is helpful to have representatives from quality assurance, production, medical services, and industrial engineering. At least one person should have a fairly comprehensive background in ergonomics. One person should take on the role of facilitator to help keep on track. In most cases, members will take on several roles. For example, the medical person will act as the facilitator, while one of the engineers may act as an ergonomics expert.

PROCEDURES

1. Meet with local management to determine the best method for ergonomics committee selection and implement that process.
2. Select committee members.
3. Hold the first meeting.
   - Establish a mission statement
   - Establish the level of commitment expected of the members
   - Become oriented to the ergonomics process
   - Practice brainstorming
   - Set a time for training

HINTS

For a team approach to be successful, it is critical that everyone feels comfortable contributing. Regardless of the rank or position of the committee members, each person's input must be taken seriously. This nonjudgmental attitude can be very difficult to achieve. This may be facilitated by having the group function in some non-work-related manner. The group may try to solve a few creative thinking exercises before they ever learn about ergonomics.

EXAMPLES

The following steps were performed by Ace Company Inc. to initiate the formation of their ergonomics committee.

- Determine group coordinator
- Decide on proactive vs. reactive approach
- Determine criteria
  - Complete criteria worksheet
  - Perform baseline criteria research
- Recruit ergonomics team members
- Train Committee Members
- Determine team model to use
- Set up list of target jobs
- Determine schedule for target jobs
- Initiate job-site analysis process

Once the committee members had been working together for a few months, they were able to accomplish quite a bit, as reflected by the minutes from a typical meeting.

Ergonomic Committee Meeting Minutes
Ace Company Inc.

Attendance

Bill Smith	Q.A. Manager
Don Harris	IE processing
Mary Scott	Line Manager
Hank Lewis	Assistant Facility Manager
John Gail	Line worker
Louise Harper	Occupational nurse
Dan Mitchel	Assembly worker
Chari Washington	Clerk
Ben Aviator	Maintenance worker

- reading of facility manager's response to last month's list of suggestion:

Once again you have provided me with valuable information for consideration. I have placed on the agenda for my next executive meeting a discussion of your suggestions for cutting down boxes in the receiving department to make it easier to reach in. As we are currently recycling our boxes, there appears to be no reason to prevent this practice.

Also regarding your suggestion stemming from your analysis of the baking area, I have instructed purchasing to look into having boxes shipped in 50-pound weights only.

Thank you again and I look forward to your next report.

- Review of last month's list of suggested interventions for the Globe Transfer Operation. Additional suggestions are added.
- Presentation of this month's target job background.
- Initial viewing of videotape of this month's target job.
- Risk identification of this month's target job.
- Intervention discovery of this month's target job.
- Adjourn.

Additional information regarding the ergonomics committee setup and how it fits into an overall ergonomics process is included in the Corporate Ergonomics Plan in Appendix A.

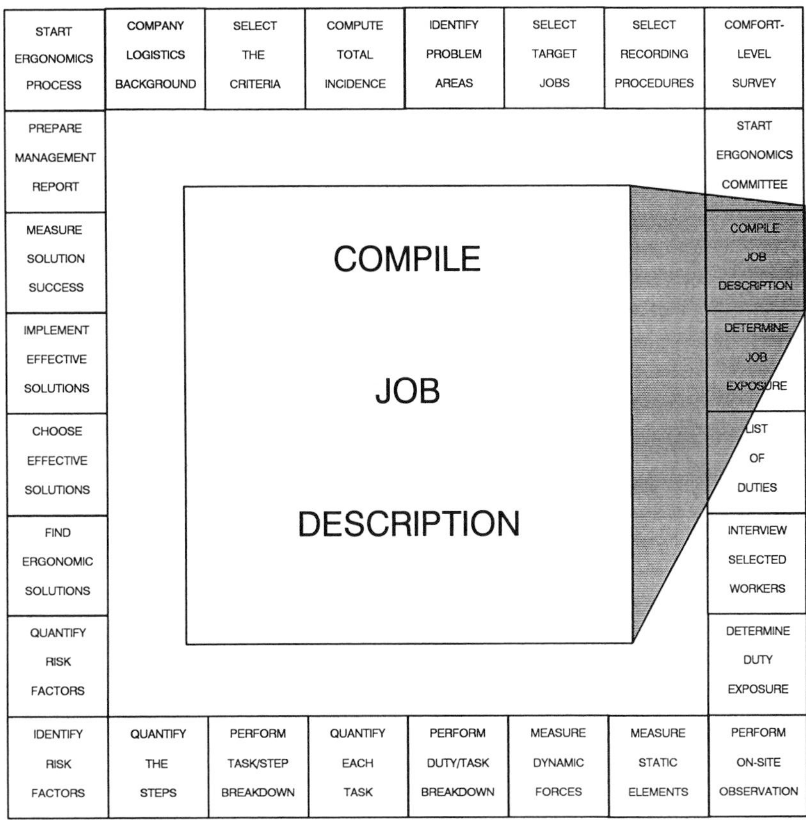

START ERGONOMICS PROCESS
COMPANY LOGISTICS BACKGROUND
SELECT THE CRITERIA
COMPUTE TOTAL INCIDENCE
IDENTIFY PROBLEM AREAS
SELECT TARGET JOBS
SELECT RECORDING PROCEDURES
COMFORT-LEVEL SURVEY
START ERGONOMICS COMMITTEE
COMPILE JOB DESCRIPTION
DETERMINE JOB EXPOSURE
LIST OF DUTIES
INTERVIEW SELECTED WORKERS
DETERMINE DUTY EXPOSURE
PERFORM ON-SITE OBSERVATION
MEASURE STATIC ELEMENTS
MEASURE DYNAMIC FORCES
PERFORM DUTY/TASK BREAKDOWN
QUANTIFY EACH TASK
PERFORM TASK/STEP BREAKDOWN
QUANTIFY THE STEPS
IDENTIFY RISK FACTORS
QUANTIFY RISK FACTORS
FIND ERGONOMIC SOLUTIONS
CHOOSE EFFECTIVE SOLUTIONS
IMPLEMENT EFFECTIVE SOLUTIONS
MEASURE SOLUTION SUCCESS
PREPARE MANAGEMENT REPORT
COMPILE JOB DESCRIPTION

Compile Job Description

OBJECTIVE

To write a job summary statement that describes the worker's function.

SUPPLIES

1. Human resource department's job description
2. Dictionary of occupational titles

GUIDELINES

This is the first of several steps in the information gathering process for your ergonomics analysis. You will write a simple paragraph that describes the worker's function. This is called the job summary statement. Within this statement you should describe the worker's function, his or her duties, and clearly define what is the overall input and output of this worker's process.

PROCEDURES

Answer the question, "What does this person do?" Try to limit this to a paragraph or two. The idea is to state the nature of this job. This step should provide you with a sense of how this job fits into the overall process of the facility. This will help to keep a perspective when you later begin to dissect the various duties, tasks, steps, or components of this job.

This may be accomplished in several ways. Most human resource departments maintain a job description for each position. Many of them already have a list of responsibilities or duties for each of their workers. This is an excellent place to *start*. These job descriptions are often fairly idealistic. There may have been some organizational changes, procedures may have changed, new equipment may have been brought in, or the company may have already instituted a job rotation system.

In some cases, the job description maintained by the employer is known to be insufficient or outdated. In those cases, the next best place to start is the dictionary of occupational titles. This book is published by the department of labor. It contains just about every job currently being performed. It can be somewhat difficult to find the target job in this book. While there are rather complete instructions in the book, often it easiest to just look up the job titles in the alphabetical index. Sometimes the same job may have several different titles. You just need to look up enough similar jobs to find it. The dictionary of occupational titles can be found at most libraries.

A third option in *starting,* if the local health and safety director or human resource is familiar enough with the jobs, is to ask this individual to provide you with information. In some cases, the on-line management can help you to develop this. This person should have first-hand knowledge of the job.

The job summary statement should cover most of this person's job description in one page. A very simple example is:

> The stamper takes brown bags from a conveyor belt, stamps them with a decorative stamp, and places them back onto the conveyor belt. On occasion this person sweeps the floor. The bags are placed on the belt by the loader and removed from the belt by the dryer loader.

As you can see, this is not completely comprehensive. It is just a way of trying, as quickly as possible, to summarize what it is that this person does and how he or she fits into the process. Another example follows.

Do not get hung up on trying to get too specific in this step. There will be time later to get verification of the various duties from the hourly worker.

Job Title *Railroad Switchman*

Date *6/12/73*

Analyst *Mike Brown*

JOB SUMMARY STATEMENT

Directs the motion of locomotive engines to move boxcars from one location to another inside a railroad switchyard.

Complete trains come into the yard and will then be broken down and placed in the appropriate storage tracks. Cars may be retrieved and placed in a specific order to be transported to another yard or local industries.

Locations and movement instructions are given on computer printouts.

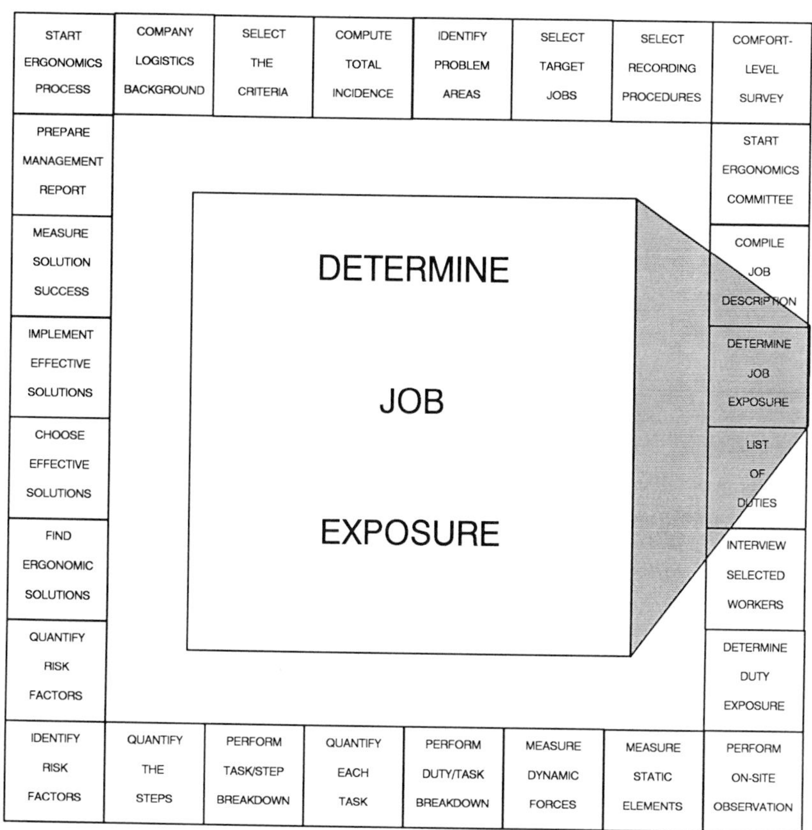
START ERGONOMICS PROCESS
COMPANY LOGISTICS BACKGROUND
SELECT THE CRITERIA
COMPUTE TOTAL INCIDENCE
IDENTIFY PROBLEM AREAS
SELECT TARGET JOBS
SELECT RECORDING PROCEDURES
COMFORT-LEVEL SURVEY
START ERGONOMICS COMMITTEE
COMPILE JOB DESCRIPTION
DETERMINE JOB EXPOSURE
LIST OF DUTIES
INTERVIEW SELECTED WORKERS
DETERMINE DUTY EXPOSURE
PERFORM ON-SITE OBSERVATION
MEASURE STATIC ELEMENTS
MEASURE DYNAMIC FORCES
PERFORM DUTY/TASK BREAKDOWN
QUANTIFY EACH TASK
PERFORM TASK/STEP BREAKDOWN
QUANTIFY THE STEPS
IDENTIFY RISK FACTORS
QUANTIFY RISK FACTORS
FIND ERGONOMIC SOLUTIONS
CHOOSE EFFECTIVE SOLUTIONS
IMPLEMENT EFFECTIVE SOLUTIONS
MEASURE SOLUTION SUCCESS
PREPARE MANAGEMENT REPORT
DETERMINE
JOB
EXPOSURE

Determine Job Exposure

OBJECTIVE

To determine the amount of time spent performing a job and to collect additional information relative to the working conditions.

SUPPLIES

1. Job exposure worksheet
2. Background information form

GUIDELINES

In this step you deal with some of the many other variables that may come into play later in the analysis process. If this is an in-house operation, then you may be familiar with this information. If you are working as an outside consultant, you will need an organized method of gathering these data. The first type of information is how much time is spent actually working at a job. Many companies provide paid lunches or breaks. This time should be deducted from the total time spent at the workplace. This adjusted total time is called the *job exposure*. The type of information in which you are interested will deal with training, promotions, meal breaks, uniforms, and other related details.

This is similar to the very first step of this process in which you recorded information about the company as a whole. The important points to cover are those that relate to the logistics and working conditions of this job. This should even include aspects such as the length of the workday, smoking policies, and supervisory responsibilities. In addition, you want to be sure that you have an accurate baseline from which to work. Be sure to note any changes that have occurred at that workstation within the past few years. These changes may be alterations to equipment, new processes, quota adjustments, or personnel reorganization.

One example of how you would utilize the type of information gathered here would be to time the length and frequency of paid breaks to determine the number of hours actually spent working in a given day. This will allow you to calculate the total number of repetitions of a task, step, or component performed each day.

PROCEDURES

1. Meet with the representative for the job being analyzed.
2. Fill out the job exposure worksheet.
3. Fill out the background information form.

The background information form can assist you in pulling together this information. Like all the forms provided here, it may need to be customized to meet particular situations.

JOB EXPOSURE WORKSHEET

Job Title	*Material handler*	Date	*7/11/90*
Job #	*9-22*	Analyst	*Offner*

Number of Shifts per Week	*5*	
Initial Daily On-Duty Time of Day	*6:30 a.m.*	*06:30*
Final Daily Off-Duty Time of Day	*4:00 p.m.*	*16:00*
TOTAL AT-WORK TIME		*9.5 hours*
Meal Break	*1 hour*	
Other Breaks	*15 minutes*	
Number of Breaks per Day	*2 coffee breaks*	
Other Non-working Time	*N.A.*	
TOTAL NON-WORK TIME		*1.5 hours*

	TOTAL AT-WORK TIME	*9.5*
Minus	TOTAL NON-WORK TIME	*- 1.5*
	JOB EXPOSURE	*8.0 hours*

BACKGROUND INFORMATION

Job Title *Egg inspector*

Date *6/17/95*

Analyst *Luke Burke*

Information Provided by *Teri Elliott*

Hours/Day *8* Days/Week *5*

Regular Hours From *8* To *5* Variable Shifts *No*

Breaks Frequency *2/day* Duration *10 min* (Regular)/Variable Control *No*

Meals Frequency *1* Duration *60 min* (Regular)/Variable Control *No*

Regulatory Hours of Service Laws *No*

Paid by (Hour) Day Piece Quota Bonus

Production Rate Required Yes (No) *But productivity monitored*

Multiple Workstations Yes (No) Workload available at shift start Yes (No)

Work Independently (With Others) Necessary Part of Crew *Dependent on delivery*

Supervised Directly Yes (No) Supervisory Duties Yes (No)

Indoors/Outdoors *Indoors*

Climate/Hazards *No*

Types of Equipment/Vehicles Used *Micrometer, scale*

Protective Equipment *Cotton gloves*

Dress Code / Uniforms *Yes* (Mandatory)/Optional

Training Necessary *Food handling standards*

Transfer (Possible) Required Promotion (Possible) Required

Union Representation *No*

Light-Duty Available *Considered light duty*

Part-time Available *No*

THE BURKE PROTOCOL F-BGINF2

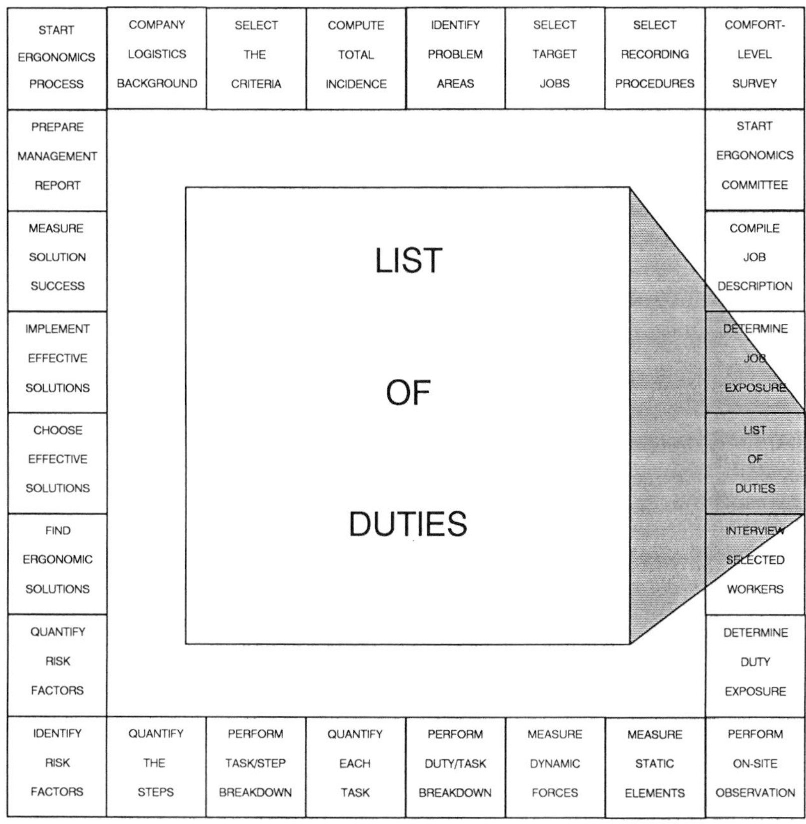
START ERGONOMICS PROCESS
COMPANY LOGISTICS BACKGROUND
SELECT THE CRITERIA
COMPUTE TOTAL INCIDENCE
IDENTIFY PROBLEM AREAS
SELECT TARGET JOBS
SELECT RECORDING PROCEDURES
COMFORT-LEVEL SURVEY
START ERGONOMICS COMMITTEE
COMPILE JOB DESCRIPTION
DETERMINE JOB EXPOSURE
LIST OF DUTIES
INTERVIEW SELECTED WORKERS
DETERMINE DUTY EXPOSURE
PERFORM ON-SITE OBSERVATION
MEASURE STATIC ELEMENTS
MEASURE DYNAMIC FORCES
PERFORM DUTY/TASK BREAKDOWN
QUANTIFY EACH TASK
PERFORM TASK/STEP BREAKDOWN
QUANTIFY THE STEPS
IDENTIFY RISK FACTORS
QUANTIFY RISK FACTORS
FIND ERGONOMIC SOLUTIONS
CHOOSE EFFECTIVE SOLUTIONS
IMPLEMENT EFFECTIVE SOLUTIONS
MEASURE SOLUTION SUCCESS
PREPARE MANAGEMENT REPORT
LIST OF DUTIES

List of Duties

OBJECTIVE

To compile a list of the duties for the job to be studied.

SUPPLIES

1. Job summary statement, and/or
2. Job description, and/or
3. Dictionary of occupational titles

GUIDELINES

The goal of this step is to compile a complete and accurate list of duties for which the target worker is responsible. A *duty* is a general statement that describes an activity. For the purposes of the ergonomic analysis, this should be an action word.

This step is essential in order to be able to later calculate the levels of exposure to each injury risk factor in your analysis. While it is not absolutely necessary to perform this prior to your observation and analysis, it is helpful to have some information beforehand to assist you in focusing your attention on certain activities.

PROCEDURES

Start with the job summary statement, a job description from human resources, or the description in the dictionary of occupational titles. Convert the information into a list. List all the duties for which the worker is responsible and assign a number to each duty. Later you will get verification of the completeness and accuracy of the list, as well as information regarding how much time is spent performing each duty. This latter step will involve verification from both on-line management and hourly workers.

The list of duties should contain action words. Rather than expressing a duty as "responsible for the overall cleanliness of the kitchen," a better statement would be "responsible for cleaning floors, tables, cooking equipment, and utensils." Another example would be stating that a secretary is "responsible for providing administrative support to the department head." It is far more useful to state such duties as "typing, filing, and talking on the phone."

Job Title *Railroad switchman*

Date *7/11/91*

Analyst *Adam Vincent*

LIST OF DUTIES

Units

# DESCRIPTION OF DUTY	Expo	%
D1- Breaks down incoming trains.		
D2- Builds outgoing trains.		
D3- Moves broken railroad cars to rip track.		
D4- Attaches cabooses to outbound trains.		
D5- Delivers boxcars to local industry.		
D6- Weighs special loads.		

THE BURKE PROTOCOL *DUTIES-1*

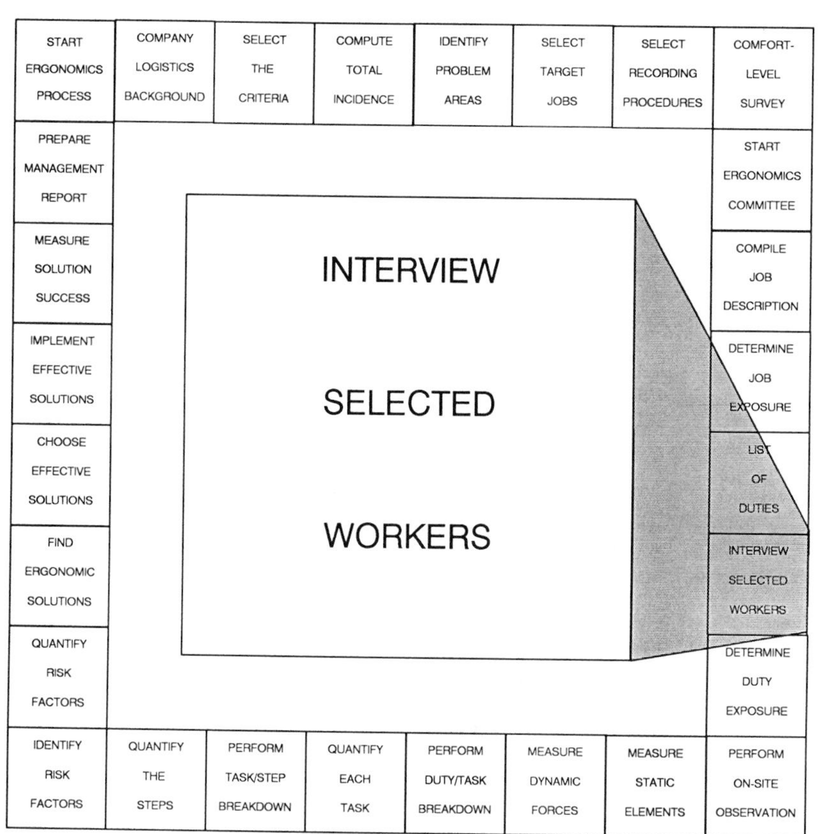
START ERGONOMICS PROCESS
COMPANY LOGISTICS BACKGROUND
SELECT THE CRITERIA
COMPUTE TOTAL INCIDENCE
IDENTIFY PROBLEM AREAS
SELECT TARGET JOBS
SELECT RECORDING PROCEDURES
COMFORT-LEVEL SURVEY
START ERGONOMICS COMMITTEE
COMPILE JOB DESCRIPTION
DETERMINE JOB EXPOSURE
LIST OF DUTIES
INTERVIEW SELECTED WORKERS
DETERMINE DUTY EXPOSURE
PERFORM ON-SITE OBSERVATION
MEASURE STATIC ELEMENTS
MEASURE DYNAMIC FORCES
PERFORM DUTY/TASK BREAKDOWN
QUANTIFY EACH TASK
PERFORM TASK/STEP BREAKDOWN
QUANTIFY THE STEPS
IDENTIFY RISK FACTORS
QUANTIFY RISK FACTORS
FIND ERGONOMIC SOLUTIONS
CHOOSE EFFECTIVE SOLUTIONS
IMPLEMENT EFFECTIVE SOLUTIONS
MEASURE SOLUTION SUCCESS
PREPARE MANAGEMENT REPORT
INTERVIEW
SELECTED
WORKERS

Interview Selected Workers

OBJECTIVE

To acquire hands-on information and verification of several aspects of the worker's duties, attitudes, and ideas for improving the workstation.

SUPPLIES

1. An open nonjudgmental attitude
2. A willingness to listen

GUIDELINES

If you have not had the opportunity to meet with hourly workers, either at an initial orientation meeting or as you were performing the data gathering, you should try to arrange such a discussion. The purpose of this meeting is to get their input as to specific concerns or suggestions they have regarding modifications to their workstations. They can also help to point out some tasks that have not been mentioned that are typically performed infrequently. You might try asking what the workers consider the most difficult or physically demanding aspect of their jobs. Be sure to explain that you are collecting suggestions for making the job easier, safer, or more efficient.

PROCEDURES

This interview process can take place as a group meeting, a formal one-on-one meeting, or as an informal chat at the workstation.

THE TYPE OF INFORMATION YOU WANT IS

1. Accuracy of the duty list
2. Input regarding duty exposure
3. Tasks that make up each duty
4. Input regarding task exposure
5. Input regarding the toughest or most physically challenging aspect of the job
6. Suggestions for making the job easier or safer
7. Any regular physical complaints as a result performing any or all tasks

This is an appropriate time to address some considerations regarding your interactions with the hourly workers. While this is usually an issue only if you are functioning as an outside consultant, even an in-house ergonomics specialist needs to be aware of some of these concerns.

Whenever you are interacting with a worker, try to remember that your priorities are not necessarily the same as his. He may suspect your intentions. If he is working at the time, he may need to pay close attention to perform a task that to an outsider does not seem that complex. It is also possible that he just does not like you.

When interviewing a worker, take as few notes as possible. This will make it seem more like a conversation than an interview or an interrogation. For the most part, people enjoy talking about their jobs, unless they think that what they say may be held against them. If possible, try to write down as much as possible immediately following talking to an employee. Like many of the skills presented in this book, with a little practice you will find that you can remember a great deal of detail.

There are some words to be avoided here. Words such as *productivity, automation, staff reduction,* or even *carpal tunnel syndrome* can lead to panic and mistrust. If you can, tell a worker that you are there to try to make their job easier, make their job station more comfortable, and make their job safer. If there is already an awareness of a high incidence of injuries, you can say that you are part of an injury-reduction program.

Examples

The following is a sample conversation that might occur if you were trying to determine the duties and steps involved in a kitchen worker's job.

Could you tell me about your job?

I set up the kitchen in the morning, cook the food at the grill, clean up as I go, stock the supplies after the lunch rush, and clean up until the night guy comes in.

What do you have to do to set up the kitchen?

Well, I have to set up the line.

What does that mean?

You know, put out the tomatoes, and the lettuce, pickles, and onions and stuff.

Where does all that come from? Does someone bring it to you or do you have to get it yourself?

No, no, I have to get it myself. From the walk-in.

Is it all ready for you? In the pans that you put on the line?

No, I have to put it all in the pans. It's in boxes and stuff.

Do you have to cook it or prepare it in any way?

No, except the tomatoes. I have to cut the tomatoes.

Okay, let me make sure I have this straight. You come in in the morning, go to the walk-in refrigerator, get out the bags and boxes of onions and pickles and stuff, get out the tomatoes, cut the tomatoes, put all this into these flat pans, and put them out on the line. Is that right?

Yeah. That's right.

What do you consider to be the most difficult part of your job?

I don't know. It's all pretty hard.

Well, if there were one thing that was most physically challenging, what would that be?

I told you I think it's all pretty hard.

In that case what do you personally hate to do?

Oh, in that case, cleaning the grill. Once a month we have to take the grill apart, carry the top outside, and hose it down.

Can you think of anything that would make it easier?

Well sure, if the night shift sprayed the grill with oven cleaner before we had to clean it then I wouldn't have to work so hard. But you'll never get them to do it.

All I can do is try.

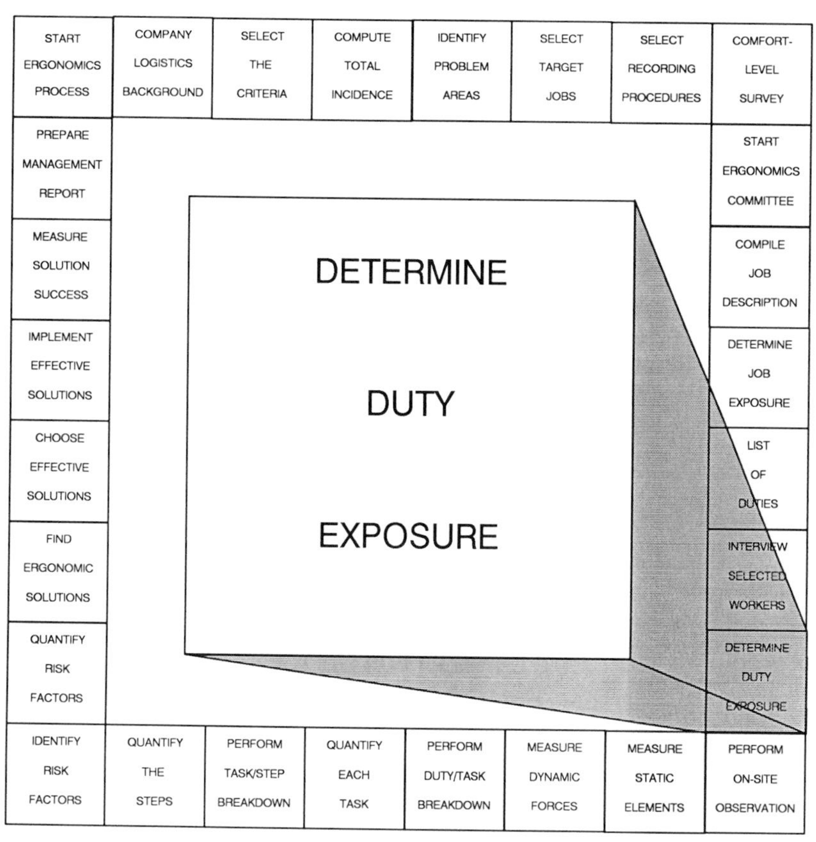
START ERGONOMICS PROCESS
COMPANY LOGISTICS BACKGROUND
SELECT THE CRITERIA
COMPUTE TOTAL INCIDENCE
IDENTIFY PROBLEM AREAS
SELECT TARGET JOBS
SELECT RECORDING PROCEDURES
COMFORT-LEVEL SURVEY
START ERGONOMICS COMMITTEE
COMPILE JOB DESCRIPTION
DETERMINE JOB EXPOSURE
LIST OF DUTIES
INTERVIEW SELECTED WORKERS
DETERMINE DUTY EXPOSURE
PERFORM ON-SITE OBSERVATION
MEASURE STATIC ELEMENTS
MEASURE DYNAMIC FORCES
PERFORM DUTY/TASK BREAKDOWN
QUANTIFY EACH TASK
PERFORM TASK/STEP BREAKDOWN
QUANTIFY THE STEPS
IDENTIFY RISK FACTORS
QUANTIFY RISK FACTORS
FIND ERGONOMIC SOLUTIONS
CHOOSE EFFECTIVE SOLUTIONS
IMPLEMENT EFFECTIVE SOLUTIONS
MEASURE SOLUTION SUCCESS
PREPARE MANAGEMENT REPORT
DETERMINE DUTY EXPOSURE

Determine Duty Exposure

OBJECTIVE

To determine the relative exposure or amount of time spent performing each duty.

GUIDELINES

In order to facilitate prioritizing the need for ergonomic intervention, it is essential to quantify the exposure to each duty. You are going to try to state how much actual time is spent at each duty. If a person performs the same routine daily, this is simple. You might be able to state that they spend $6^1/_2$ out of 8 hours operating the crank stamping machine, 45 minutes on preventative maintenance, 30 minutes cleaning up, and 15 minutes on paperwork.

Not all jobs are that regular. In some cases, the person performs different duties for varying amounts of time every day. You will have no choice but to try to average out the different duties. An example is "Worker spends 2 hours driving to the post office twice per month." This means that he/she spends 4 hours per month driving to the post office. Four hours per month averages out to 1 hour per week, or even 12 minutes per day. This is obviously an artificial number. This breakdown is only used for the purpose of comparing the exposure to this duty to exposure to other duties that this worker performs. In the final report, if this duty is analyzed, it will be referred to by its actual time per month format.

Once all the duties have been reduced to hours per day or percents, you can determine the relative exposure level. This is important because the figures derived here will allow you to compute the relative exposure levels for the many risk factors as they are identified.

The next step is to compute the percent of total time spent performing a duty. As the units are hours per day, then the percentage would be the number of hours, or fraction thereof, that each duty is performed divided by the total daily job exposure in hours times 100.

List of Duties — Physical Therapist

Duty	Exposure	Common Unit	Percent
Paperwork	1 hr/day	1.0 hr/day	13.5
Patient care	$5^1/_2$ hr/day	5.5 hr/day	74.3
Talking on phone	30 min/day	0.5 hr/day	6.8
Cleaning up	20 min/day	0.3 hr/day	4.1
Equipment maintenance	2 hr/month	0.1 hr/day	1.4
Total Daily Exposure		7.4 hr/day	

PROCEDURES

Before you can perform all the calculations above, you must obtain information about the accuracy of your list of duties and the amount of time spent on each one. Just about the only source for this kind of information is the hourly worker and the on-line supervisor.

One approach is to print out the list, double-spaced with large margins. Make several copies. Start with the on-line manager. Ask him/her to review the list, make any deletions, additions, or modifications. Also ask him/her to enter how much of the worker's time is spent performing each task. Next, spend a little time with several workers and ask them to perform the same review and editing. Check with as many workers as is practical. Be sure to give each worker his or her own copy. The last thing you want to do is get two or three workers arguing over how to fill out the sheet.

Often you will encounter someone who will tell you that his job cannot be reduced to a list of duties because it varies too much from day to day. In those cases you must ask him about the average day. The worker will likely object to the concept of the "average" day. He will say that there is no such thing as an average day. In those cases, you can expand the question to the average month. In some cases you will have to talk in terms of the average year or the average 20 years. You may need to tell him to consider every day that he has worked there for the past 20 years, and if he has to account for all that time, how would it break down into the number of times he has performed a certain duty.

This is not an easy process for the worker. One way you can assist him is to give him parameters. Ask him, "In an average day would you say you perform this duty closer to 5 minutes or closer to 6 hours?". If he responds, "Closer to 5 minutes," then ask, "Is it closer to 5 minutes or 60 minutes?" If he responds, "Sixty minutes," ask, "Is it closer to 1 hour or 3 hours?". This type of exploration may be necessary, using many different time frames.

Once you have this information, return to the duties list. Fill in the exposure in the "Expo" column using a common time increment, such as hours per day, hours per week, or hours per month. Next fill in the percent of time performing each duty in the "%" column.

There is one other factor to consider. Not everyone can read. Not everyone speaks English. Just because they don't have what you personally consider minimal communication skills does not mean they are stupid. Treat them with respect, and you will get more cooperation than you ever thought possible. This is also an opportunity for them to ask you for information and advice. Be open, honest, and completely objective. Do not make any statements regarding the potential hazards of the workstation or their job. The example that follows may be helpful.

Job Title *Railroad switchman*

Date *7/11/91*

Analyst *Adam Vincent*

LIST OF DUTIES

Units *Hours*

# DESCRIPTION OF DUTY	Expo	%
D1- Breaks down incoming trains.	*2.75*	*35*
D2- Builds outgoing trains.	*3.00*	*39*
D3- Moves broken railroad cars to rip track.	*.50*	*6*
D4- Attaches cabooses to outbound trains.	*.25*	*3*
D5- Delivers boxcars to local industry.	*1.00*	*13*
D6- Weighs special loads.	*.25*	*3*

THE BURKE PROTOCOL *DUTIES-1*

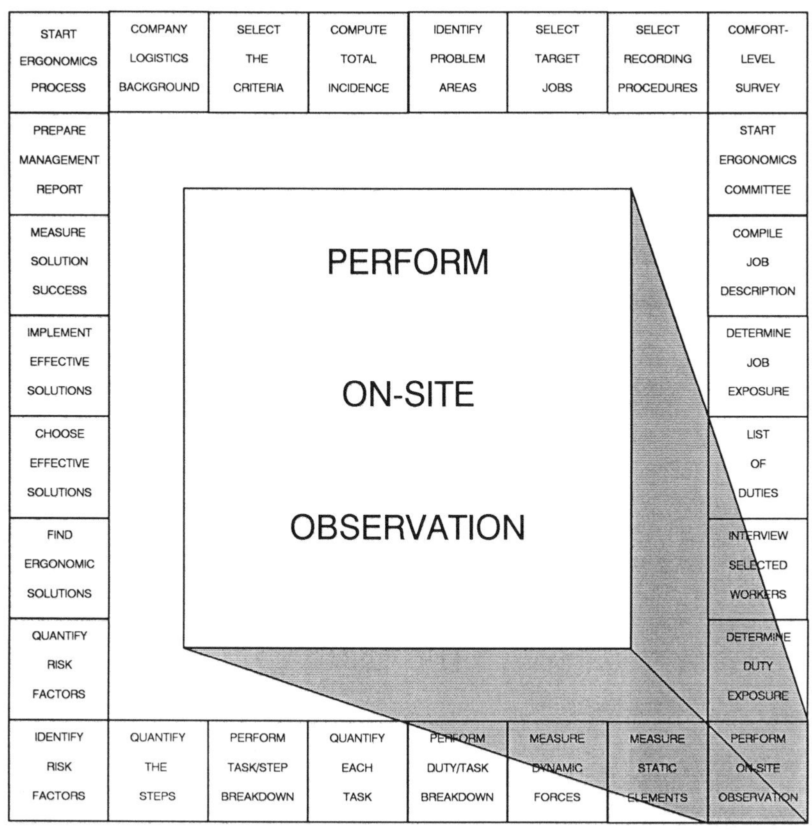

START ERGONOMICS PROCESS
COMPANY LOGISTICS BACKGROUND
SELECT THE CRITERIA
COMPUTE TOTAL INCIDENCE
IDENTIFY PROBLEM AREAS
SELECT TARGET JOBS
SELECT RECORDING PROCEDURES
COMFORT-LEVEL SURVEY
START ERGONOMICS COMMITTEE
COMPILE JOB DESCRIPTION
DETERMINE JOB EXPOSURE
LIST OF DUTIES
INTERVIEW SELECTED WORKERS
DETERMINE DUTY EXPOSURE
PERFORM ON-SITE OBSERVATION
MEASURE STATIC ELEMENTS
MEASURE DYNAMIC FORCES
PERFORM DUTY/TASK BREAKDOWN
QUANTIFY EACH TASK
PERFORM TASK/STEP BREAKDOWN
QUANTIFY THE STEPS
IDENTIFY RISK FACTORS
QUANTIFY RISK FACTORS
FIND ERGONOMIC SOLUTIONS
CHOOSE EFFECTIVE SOLUTIONS
IMPLEMENT EFFECTIVE SOLUTIONS
MEASURE SOLUTION SUCCESS
PREPARE MANAGEMENT REPORT
PERFORM
ON-SITE
OBSERVATION

Perform On-Site Observation

OBJECTIVE

To observe and, if possible, videotape the various duties and tasks being performed.

SUPPLIES

1. Video camera
2. Videotape
3. List of duties

The main purpose of this step is to learn as much as you can about the work that an individual does. It is not as important to understand all the details about the process, or how the machinery works, as it is to become fully acquainted with what the worker does.

This on-site time is extremely valuable. Focus on the goal of information gathering pertaining to the worker. Many well-meaning persons will try to explain the mechanical, chemical, or thermal reactions in detail. This is important to them and it is what they know best, but it is probably not relevant to your goal. If it becomes important, that information can be attained at a later date.

This observation may be accomplished in person, by viewing a videotape, or by a combination of both techniques. By far the best method is to observe the worker in person and then operate the video camera yourself. This will allow you to use the video camera as a note-taking device. It is important to remember that the more videotape you shoot, the more you will have to watch.

In some cases, 10 minutes of a worker feeding plastic bags into a printing machine will provide no more data than 30 seconds. In other cases, over an hour of workers performing a fast-paced highly repetitious assembly operation will not seem to be sufficient. The amount of shooting you do will vary on the complexity of the job, your experience, and your subject's response.

Most people will act differently if they know they are being videotaped. Try to perform observation of a task as unobtrusively as possible before starting to videotape. This will allow you to be sensitive to any radical changes in behavior once you turn on the camera. It may be necessary to pretend to shoot videotape for a while just to get workers used to the idea of being taped.

There are some situations in which videotaping is prohibited. In those cases, plan to spend a good deal of time just watching. Do no more than one job at a time using this method. This will allow you to get a picture in your memory that you can review when you are working on your report. You should then immediately perform the task breakdown and ergonomic risk identification outlined in the next chapters.

There are also some situations in which it may be impossible to observe a person at all. This may be due to a potentially hazardous or sensitive operation where the risk to you or to the process is too great. In that case, someone else may do the videotaping for you. If at all possible, try to give some instructions to that individual regarding the type of information you would like recorded.

Another situation where you may not be able to perform direct observation is if you are performing an analysis of seasonal work in the wrong season. For instance, if you are trying to perform an analysis of a highway maintenance man from upstate New York in June, you would have to watch any available videotape of snow plowing or other winter activities.

PROCEDURES

Always ask permission of the individual you are about to videotape. Even though you may have received an official company permission, it is just common courtesy. In this way the worker understands that this is to analyze the job and not to criticize him personally.

Try to shoot an establishing shot first. This will be somewhat far away and will show as much of the surrounding area as is practical. If at all possible, shoot this from in front and then about 90° to the side. About 30 seconds of each should be sufficient. Next, shoot the worker in a full shot. That means that his entire body takes up most of the frame. Once again, shoot from at least two adjacent angles.

If there is a specific concern, such as arm or wrist motions, you may want to shoot some additional tape of that body part. In most cases you will never want to show less than one quarter of a person's body. In other words, if you want to focus on the wrist motion, try to have the arm and half of the torso in the frame as well. If you get too close, it will be difficult for anyone who was not there at the time to understand what is going on in the tape.

Don't be concerned if you cannot follow these suggestions exactly. Very few workstations lend themselves to an unobstructed view. Try not to shoot more than 10–15 minutes of footage on any tape. The cost of additional tape is worth the time you will spend rewinding and fast forwarding the tape.

Once you have completed the videotaping, give yourself the time to write down any thoughts about the process itself, but do not try to write up a final report. The types of things you might write down are subtle risk factors you noted and any ideas you had about how to improve the workstation. Then put these notes away and refer to them only after you have completed the appropriate steps. These initial impressions can be some of your best ideas, but if you allow yourself to dwell on them you will find it difficult to consider the less dramatic but equally important observations and intervention ideas.

Next, rewind your videotape to the beginning. Reset the tape counter to 000. Watch the videotape and write down the tape-counter reading for the various jobs or duties you have observed. You will probably need to pause the tape frequently. These tape counters are not absolutely accurate, but they will give you a close approximation of the location of the various parts of the tape. You cannot always tape or observe the different steps in a process in the order that they occur in real life. It is surprising how quickly you can forget the order in which you taped or observed many jobs.

Videotape Record

Company Name	Maintenance Company
Date	June 3, 1987
Analyst	Burke
Tape Counter	Description
0 to 200	Can-crushing operation
200 to 350	Garbage disposal maintenance
350 to 475	Lawn mowing
475 to 600	Cafeteria cashier
600 to 725	Furniture moving

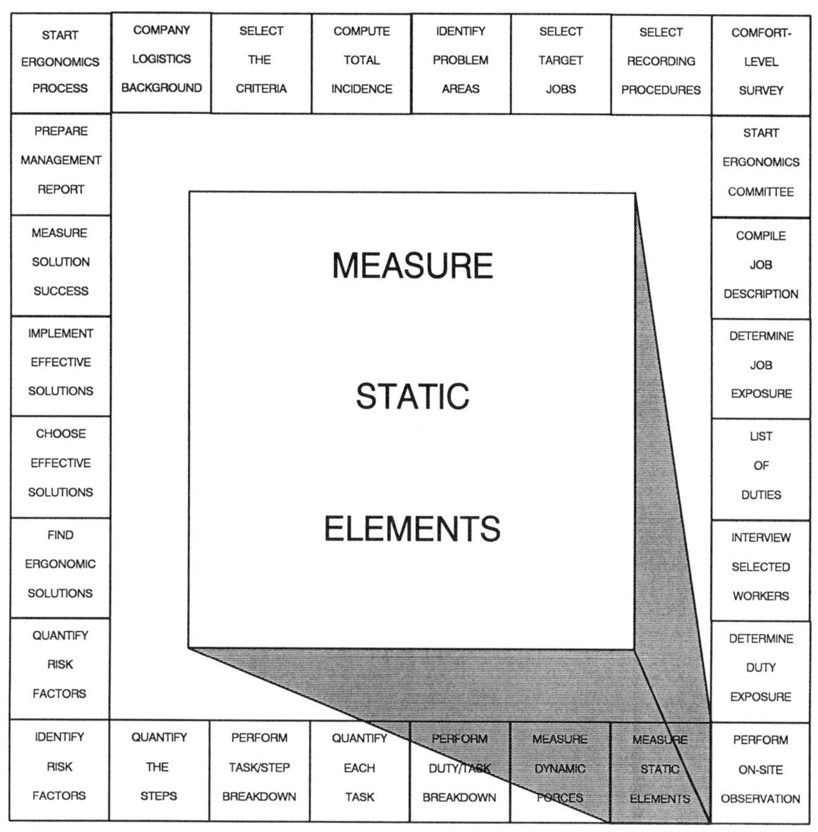
START ERGONOMICS PROCESS
COMPANY LOGISTICS BACKGROUND
SELECT THE CRITERIA
COMPUTE TOTAL INCIDENCE
IDENTIFY PROBLEM AREAS
SELECT TARGET JOBS
SELECT RECORDING PROCEDURES
COMFORT-LEVEL SURVEY
START ERGONOMICS COMMITTEE
COMPILE JOB DESCRIPTION
DETERMINE JOB EXPOSURE
LIST OF DUTIES
INTERVIEW SELECTED WORKERS
DETERMINE DUTY EXPOSURE
PERFORM ON-SITE OBSERVATION
MEASURE STATIC ELEMENTS
MEASURE DYNAMIC FORCES
PERFORM DUTY/TASK BREAKDOWN
QUANTIFY EACH TASK
PERFORM TASK/STEP BREAKDOWN
QUANTIFY THE STEPS
IDENTIFY RISK FACTORS
QUANTIFY RISK FACTORS
FIND ERGONOMIC SOLUTIONS
CHOOSE EFFECTIVE SOLUTIONS
IMPLEMENT EFFECTIVE SOLUTIONS
MEASURE SOLUTION SUCCESS
PREPARE MANAGEMENT REPORT
MEASURE STATIC ELEMENTS

Measure Static Elements

OBJECTIVE

To compile a body of information relative to the physical environment in which the worker functions.

SUPPLIES

1. Polaroid camera
2. Polaroid film
3. Graph paper (optional)
4. Straight edge
5. Calipers (diameters)
6. Straight folding ruler (depths)
7. 50-foot 3/4-inch tape measure (distances)
8. Cloth tape measure (circumferences)
9. Cloth for wiping off hands
10. Gloves

GUIDELINES

In this part you objectively measure the static elements of the workstation. This includes work heights, reach distances, walking distances, overhead or under-desk clearances, and any other pertinent data relative to the physical environment. The goal of this step is to obtain workstation design properties. You do not have the worker present when you perform this. This can be done when the worker is on break or at lunch, and even before you have ever observed this operation.

PROCEDURES

The first thing to do is to draw a sketch of the area. You do not have to be a graphic artist. These notes are for your eyes only. It is helpful to use graph paper, but only as a general guide. Unless you have the ability and experience, do not try to draw in perspective or try to stick to any rigid scale. This is a simple note-taking process to get the relative positions of objects.

Your approach is similar to the videotaping approach in that you want to draw a picture from two sides. To start, draw a two-dimensional picture from one side. Next, pick a single reference point and measure everything from that point. Measure every height, depth, and distance at the workstation. Then move

90° and repeat the process. If equipment is adjustable (such as chair height), record the high and low end ranges.

It is very important to pick one spot as a reference point and to measure everything from that spot. It is usually easiest to pick a spot on the floor. In this way, for example, each shelf is measured from the floor up. This is far easier and less confusing than measuring the distance from one shelf to another and trying to note this on the drawing.

Use the straight folding ruler to measure heights or depths, the 50-foot-3/4 inch tape measure to measure walking distances, and the cloth tape measure to measure circumferences.

There is an easier, softer way to do this job. If it is permitted, take a few instant photographs and then write the measurements directly on the picture. This eliminates any drawing, but it can be a little more difficult to fit all your measurements on the photo. Stick to whole numbers. One-half inch or 3/4-inch measurements are not usually important.

WORKSTATION SKETCH

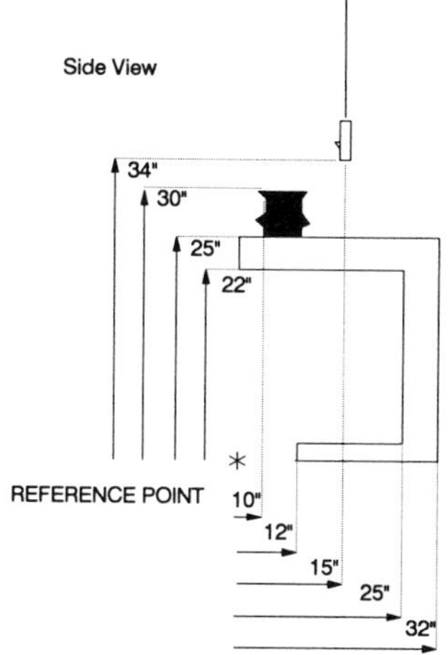

Pick a single reference point.(*)
Measure everything from that point.

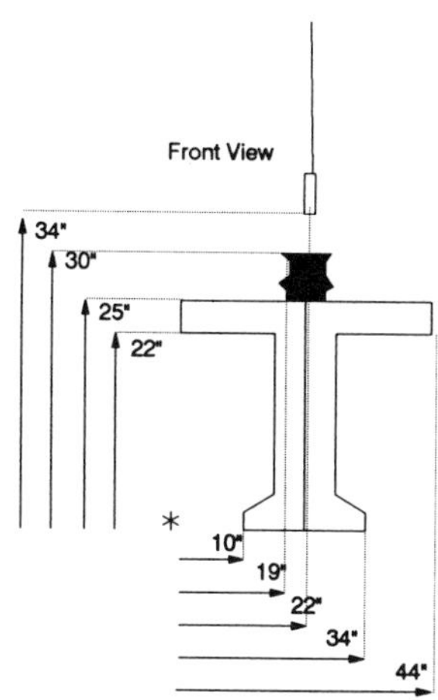

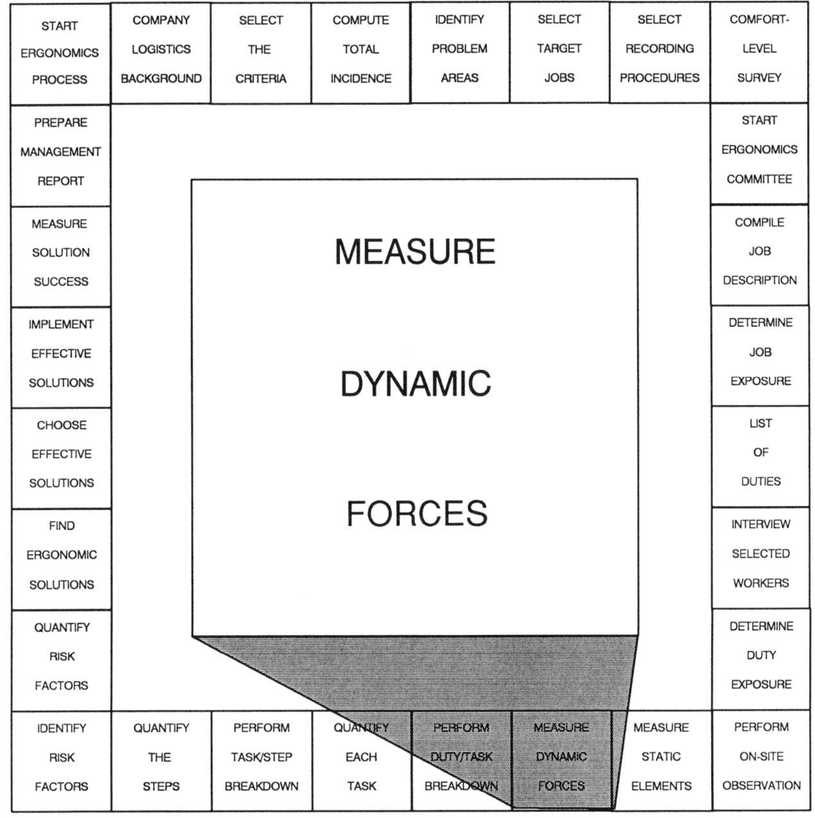

START ERGONOMICS PROCESS
COMPANY LOGISTICS BACKGROUND
SELECT THE CRITERIA
COMPUTE TOTAL INCIDENCE
IDENTIFY PROBLEM AREAS
SELECT TARGET JOBS
SELECT RECORDING PROCEDURES
COMFORT-LEVEL SURVEY
START ERGONOMICS COMMITTEE
COMPILE JOB DESCRIPTION
DETERMINE JOB EXPOSURE
LIST OF DUTIES
INTERVIEW SELECTED WORKERS
DETERMINE DUTY EXPOSURE
PERFORM ON-SITE OBSERVATION
MEASURE STATIC ELEMENTS
MEASURE DYNAMIC FORCES
PERFORM DUTY/TASK BREAKDOWN
QUANTIFY EACH TASK
PERFORM TASK/STEP BREAKDOWN
QUANTIFY THE STEPS
IDENTIFY RISK FACTORS
QUANTIFY RISK FACTORS
FIND ERGONOMIC SOLUTIONS
CHOOSE EFFECTIVE SOLUTIONS
IMPLEMENT EFFECTIVE SOLUTIONS
MEASURE SOLUTION SUCCESS
PREPARE MANAGEMENT REPORT
MEASURE DYNAMIC FORCES

Measure Dynamic Forces

OBJECTIVE

To compile a body of information relative to the physical environment in which the worker functions, focusing on the dynamic forces that must be overcome.

SUPPLIES

1. Force gauge dynamometer
2. Attachment devices for force gauge dynamometer
3. Weight scale

GUIDELINES

Based on your job summary statement and direct observation, you will measure the dynamic forces associated with the performance of this job. This includes weights, pushing and pulling forces, and torques necessary to perform a task.

To perform this you can weigh objects, look on packages for a weight, or ask someone who has weighed the object at some time in the past. If you weigh something, be sure to weigh it several times to ensure accurate and consistent results. To measure pushing and pulling forces will require a force gauge dynamometer. This is a hand-held device that acts like a scale to measure pushing or pulling forces in pounds or kilos. The trickiest part of this operation is trying to attach the force gauge in a way that closely simulates the normal hand or body part coupling with the object.

For example, if a person usually opens a door by pulling on the handle with his hand, then you need to attach the force gauge dynamometer to the door handle and try to duplicate the angle at which he applied the force (Figure 1). This may require a variety of clamping devices. Some of the devices are vice grips, wood clamps, paper clips, and chains and cables. Needless to say, this can be a very cumbersome group of tools to carry around. You can probably get by with a small section of decorative fish net and a ball of strong twine. These items seem to solve most normal coupling needs.

To measure the pushing force, line up the angle of force as closely as possible and push. You will discover that you will be concentrating your effort in an awkward position. Don't be surprised if it is very difficult for you to do something as simple as opening a door and yet the force gauge dynamometer does not read out all that high.

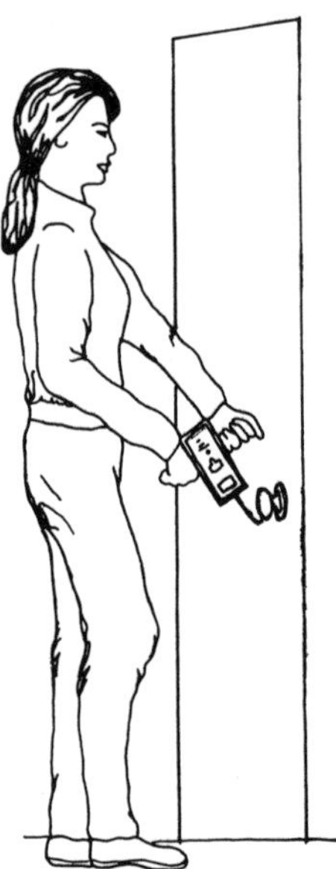

Figure 1. Using a force gauge dynamometer to measure the dynamic force required to open a door.

As with most skills, using a force gauge dynamometer requires practice. Try practicing pushing and pulling objects at work and at home. Your spouse will be thrilled to know it take 14 pounds of force to open the refrigerator door.

If no scale is available, the force gauge dynamometer can be used to measure weights. Place objects in the fish net and lift. This will measure the force of gravity pulling down on the load. **Be very careful.** You will need to hold the device a sufficient distance from your body to allow it to hang freely. This may require you to use potentially terrible body mechanics and run the risk of injuring your back. This can be embarrassing, not to mention painful.

There is a great deal of ongoing development of tools to assist in the gathering of this type of information. In particular, many computer programs exist to calculate the various forces exerted by different body parts during certain tasks. Most of these programs are strongly based on university-based objective research. In some cases, this computer-calculated information may prove valuable. In many cases, the raw information you gather using limited tools is easier to use and much more practical.

Remember to take several measurements until you get a consistent reading. This objective information will become critical in the job-analysis intervention step. There is nothing worse than being back in your office and realizing that you forgot to weigh the shovel that a worker uses to dig a ditch. It makes it very difficult to suggest that a company increase the size or weight of the shovel without that information.

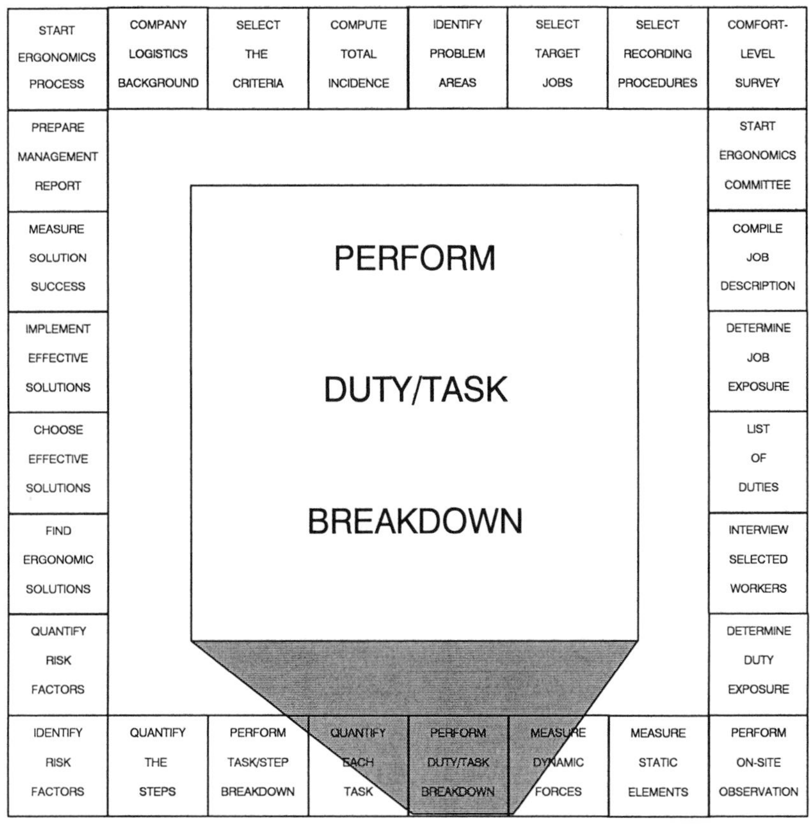
START ERGONOMICS PROCESS
COMPANY LOGISTICS BACKGROUND
SELECT THE CRITERIA
COMPUTE TOTAL INCIDENCE
IDENTIFY PROBLEM AREAS
SELECT TARGET JOBS
SELECT RECORDING PROCEDURES
COMFORT-LEVEL SURVEY
START ERGONOMICS COMMITTEE
COMPILE JOB DESCRIPTION
DETERMINE JOB EXPOSURE
LIST OF DUTIES
INTERVIEW SELECTED WORKERS
DETERMINE DUTY EXPOSURE
PERFORM ON-SITE OBSERVATION
MEASURE STATIC ELEMENTS
MEASURE DYNAMIC FORCES
PERFORM DUTY/TASK BREAKDOWN
QUANTIFY EACH TASK
PERFORM TASK/STEP BREAKDOWN
QUANTIFY THE STEPS
IDENTIFY RISK FACTORS
QUANTIFY RISK FACTORS
FIND ERGONOMIC SOLUTIONS
CHOOSE EFFECTIVE SOLUTIONS
IMPLEMENT EFFECTIVE SOLUTIONS
MEASURE SOLUTION SUCCESS
PREPARE MANAGEMENT REPORT
PERFORM
DUTY/TASK
BREAKDOWN

Perform Duty/Task Breakdown

OBJECTIVE

To break the job down into units that will allow easier identification and quantification of risk factors.

SUPPLIES

1. Videotape player with a pause feature.
2. Videotape of worker.

The job of identifying all the potentially hazardous conditions at the worksite can be overwhelming if not approached in an organized manner. This may involve breaking the duties down into tasks, steps, and components. The relationship between duties, tasks, steps, and components is a simple one.

A *duty* is a general statement of responsibility that involves a specific action or actions. Each duty is made up of one or more tasks.

Tasks are specific activities of a definable duration. Examples of tasks are "sweeps floor," "moves boxes," "types letters." In most cases, tasks are repeated several times to complete or perform a duty.

Each task may have one or more steps. A *step* is a specific action that is required to perform a task. A step must start with an action word. That action word should be taken from this list below.

Stand	Finger
Sit	Walking
Carry	Lift
Pull	Push
Balance	Climb
Bend/stoop	Kneel
Crouch	Crawl
Reach	Handle

The Duty-Task-Step Flowchart demonstrates this relationship.

A list of duties has already been prepared. The next step is to break down the duties that were observed into tasks. A task is simply a definable unit of activity that goes into the performance of a duty. This breakdown can take place either as the job is being observed or as the videotape is being reviewed. The videotape review is preferable.

Start by observing the worker performing a specific duty. Whatever specific action the worker is doing is considered the first task of that duty. That task will continue until one of the following occurs:

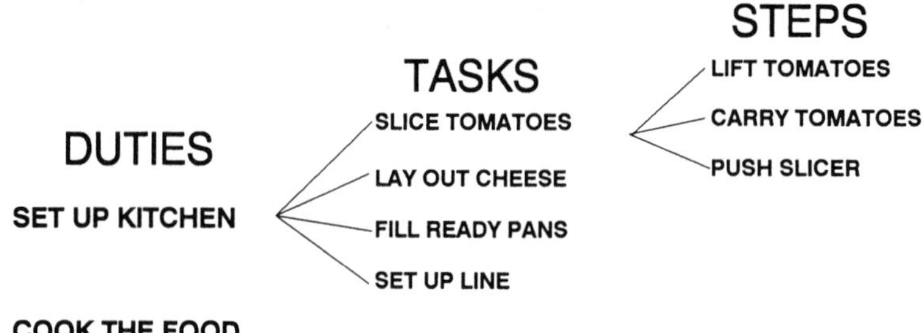

COOK THE FOOD

KEEP KITCHEN CLEAN

FAST-FOOD WORKER

RESTOCK STATION

Job Title	*Cook III*
Date	*1-16-75*
Analyst	*Henry Pitkin*

DUTY - TASK BREAKDOWN

Duty: *D1: Sets up kitchen*
- Tasks: *T1: Slices tomatoes*
  - *T2: Lays out cheese*
  - *T3: Fills ready pans*
  - *T4: Sets up line*

D2: Cooks food
- *T1: Cooks grilled sandwiches*
- *T2: Makes salads*
- *T3: Prepares drinks*

D3: Keeps kitchen clean
- *T1: Wipes down prep space*
- *T2: Washes utensils*
- *T3: Scrapes grill*

D4: Restocks station
- *T1: Replenishes ready pans on line*
- *T2: Replenishes paper plates*

THE BURKE PROTOCOL

The output (product) is changed.

The action of the worker differs.

The workstation or location changes.

The following example may help to clarify this.

THE EGG INSPECTOR

A worker is responsible for inspecting eggs before they are shipped. The egg inspector has the egg crates fed to her on the left on a gravity-fed ramp. The egg inspector removes each egg from the carton, holds it up to an inspection light, measures it with a length gauge, and replaces it in the egg carton. When 10 of the egg cartons have been completed, the inspector takes the cartons and places them in a shipping box located on the floor behind her. When five boxes are piled up behind her, she gets a handtruck and moves them to the shipping area. The egg inspector then returns to her workstation and starts the process over.

At what points did the output change?

Task 1 — Output: Cartons of Inspected Eggs

The egg inspector has the egg crates fed to her on the left from a gravity-fed ramp. The egg inspector removes each egg from the carton, holds it up to an inspection light, measures it with a length gauge, and replaces it in the egg carton.

Task 2 — Output: Boxes of Cartons

When 10 of the egg cartons have been prepared, the inspector takes the cartons and places them in a shipping box located on the floor behind her. When five boxes are piled up behind her, she gets a handtruck and moves them to the shipping area. The egg inspector then returns to her workstation and starts the process over.

Using this first criteria, this job can be broken up into two tasks.

Are the actions of the worker significantly different?

In task 1, the egg inspector performs mostly the same types of activities. She handles small items using mostly her upper body. Therefore, there is no need to break this into any additional tasks.

In task 2, there is a significant change in the types of activities being performed. Larger boxes are handled and different equipment is used. So you can break this up into two different tasks

Task 2 (revised): When 10 of the egg cartons have been completed, the inspector takes the cartons and places them in a shipping box located on the floor behind her.

Task 3 (revised): When five boxes are piled up, the egg inspector gets a handtruck and moves them to the shipping area. The egg inspector then returns to her workstation and starts the process over.

Does the workstation or location change?

If you had not already split task 2 into two different tasks, it is apparent that this final criteria would have facilitated that decision.

Therefore, the tasks may be listed as follows:

Task	Units of Completion
T1: Inspecting eggs	Cartons of eggs
T2: Loading boxes	Boxes of egg cartons
T3: Transporting boxes	Handtruck loads (5 boxes)

This procedure will become clearer as you gain experience. Also, the approach is not terribly rigid. If two people choose to break a job down into tasks differently, it will not affect the end product, as long as the relationships between the tasks and the duties are maintained. The overall purpose is to facilitate the risk identification process by grouping several activities together in some logical and practical manner.

A task is simply a definable unit of activity that goes into the performance of a duty. How vigorously you choose to break that task down will depend on experience and the time available to perform the breakdown. The purpose here is only to provide a unit of activity to be observed so that specific risk factors can be identified.

Examples follow that may prove helpful.

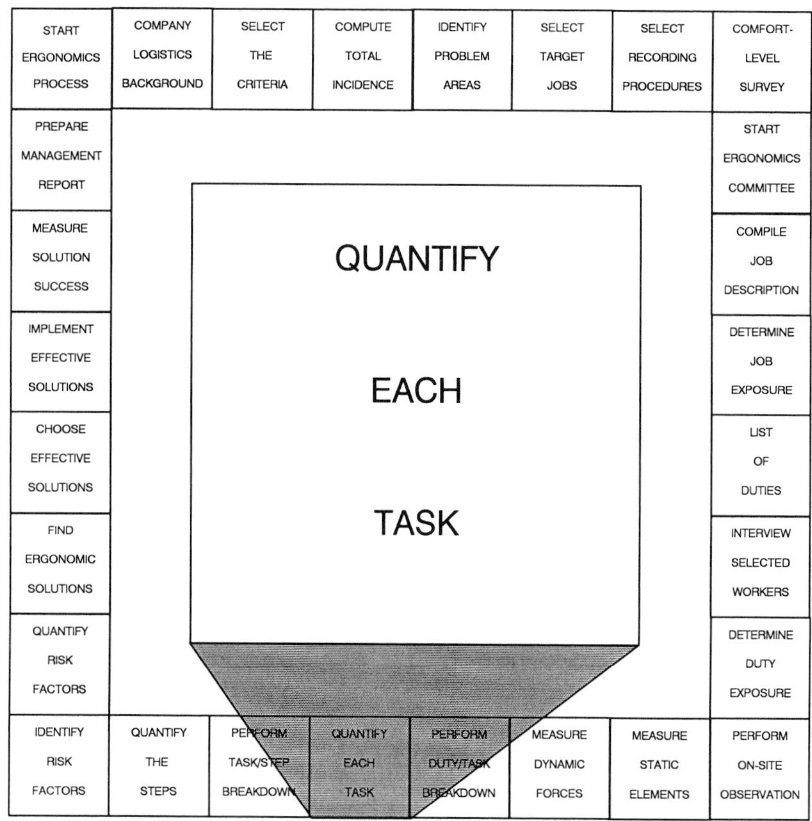

START ERGONOMICS PROCESS
COMPANY LOGISTICS BACKGROUND
SELECT THE CRITERIA
COMPUTE TOTAL INCIDENCE
IDENTIFY PROBLEM AREAS
SELECT TARGET JOBS
SELECT RECORDING PROCEDURES
COMFORT-LEVEL SURVEY
START ERGONOMICS COMMITTEE
COMPILE JOB DESCRIPTION
DETERMINE JOB EXPOSURE
LIST OF DUTIES
INTERVIEW SELECTED WORKERS
DETERMINE DUTY EXPOSURE
PERFORM ON-SITE OBSERVATION
MEASURE STATIC ELEMENTS
MEASURE DYNAMIC FORCES
PERFORM DUTY/TASK BREAKDOWN
QUANTIFY EACH TASK
PERFORM TASK/STEP BREAKDOWN
QUANTIFY THE STEPS
IDENTIFY RISK FACTORS
QUANTIFY RISK FACTORS
FIND ERGONOMIC SOLUTIONS
CHOOSE EFFECTIVE SOLUTIONS
IMPLEMENT EFFECTIVE SOLUTIONS
MEASURE SOLUTION SUCCESS
PREPARE MANAGEMENT REPORT
QUANTIFY EACH TASK

Quantify Each Task

OBJECTIVE

To determine the level of exposure for each task or duty as expressed in time (duration), frequency (rate), and total number of repetitions per day (quantity)

SUPPLIES

1. Stopwatch.
2. Videotape player with pause feature.

This step is also a preparation for the risk identification. In this step you will record the relationships between the duties and tasks. This will provide the information to quantify the risk factors once they are identified.

The data to be calculated involve defining a unit of completion, determining how long it takes to perform that unit of completion, verifying how long this task is performed in terms of time, and verifying how many times this task is performed. This information will be used to help prioritize the exposure for each task and therefore to direct the risk identification process. It will also be used in the final report to help point out the need for interventions in various areas.

PROCEDURES

The questions to ask to facilitate this process are

1. What is the unit of completion? (Unit of completion)

Express the task as a unit of completion, for instance, garments completed, pallets loaded, or pieces deburred.

2. How long does it take to perform one unit of completion? (Time to perform)

Determine the time to perform each task. This may be done by timing how long it takes to complete the task once. If a task is being performed at high speed, it may be easier to time how long it takes to perform this task several times in a row and then divide by the number of repetitions.

3. At what rate is this performed? (Rate)

Use the duration-to-rate conversion charts at the end of this chapter to convert to the time to perform to a rate. The rate at which a task, step, or component is performed is expressed in repetitions per hour.

4. How long each day does the worker perform this task? (Exposure)

Determine the amount of time in hours or fractions of hours that a worker performs this specific task. As in the case of calculating the duties exposure, this unit of time should be converted to hours per day, even if it is an artificial unit.

5. How many times does the worker perform this task each day? (Quantity)

Multiply the exposure expressed as hours per day by the rate expressed as units of completion per day to compute the quantity.

This step is similar to the task breakdown in that it is much easier to understand after you have performed the process. If you initially use the techniques described here, you will find that it is a simple process. In time you will learn to abbreviate these protocols.

The following exercises should serve as a review, as well as show how these latest steps fit into the process.

EXERCISE 1

John Smith is a warehouse worker. He works from 7:30 a.m. to 5:00 p.m. He takes two paid 15-minute coffee breaks and an unpaid 1-hr lunch. He spends about 5 hr per day loading boxes. He also spends 1 hr per day performing maintenance and cleanup. The rest of his time, about $1^1/_2$ hr, is spent putting away stock. About two times each day he does 15 minutes of paperwork.

Part I — List of Duties and Duty Exposure

Questions

1.1 What is his list of duties?
1.2 What is his total job exposure?
1.3 What is his duties exposure?

Answers

1.1 What is his list of duties?

D1 Loading boxes
D2 Maintenance and cleanup
D3 Putting away stock
D4 Performing paperwork

1.2 What is his total job exposure?

	7:30 to 5:00 =	9.5 h
Minus	Two 15-min. breaks	0.5 h
Minus	Lunch	1.0 h
	Total job exposure	8.0 h

1.3 What is his duties exposure?

D1 Loading Boxes	5.0 h/d
D2 Maintenance and cleanup	1.0 h/d
D3 Putting stock away	1.5 h/d
D4 Performing paperwork	0.5 h/d
Other	
Total duties exposure	8.0 h

Part II — Task Breakdown Quantification

Following a review of the videotape taken during the on-site observation, D1, "loading boxes," can be broken down into specific tasks.

D1. Loading Boxes

T1 Palletizing boxes from conveyor belt
T2 Transporting to shrink-wrapping area
T3 Operating shrink-wrapping machine
T4 Transporting to the loading area

Questions

1.4 What is the unit of completion for each of the tasks above?

Answers

1.4 Units of completion

T1 Completed pallet
T2 Trip to shrink wrapper
T3 Shrink wrapping a pallet
T4 Trip to loading area

During the videotape review, each task is timed.

T1	Palletizing boxes from conveyor belt	5 min
T2	Transporting to shrink-wrapping area	4 min
T3	Operating shrink-wrapping machine	3 min
T4	Transporting to the loading area	1 min

Questions

1.5 How long does it take to perform one unit of completion? (Time to perform)
1.6 At what rate is this performed? (Rate)
1.7 How long each day does the worker perform each task? (Exposure)
1.8 How many times does the worker perform this task each day? (Quantity)

Answers

1.5 How long does it take to perform one unit of completion? (Time to perform)

T1	Completed pallet	5 min
T2	Trip to shrink wrapper	3 min
T3	Shrink wrapping a pallet	4 min
T4	Trip to loading area	3 min

1.6 At what rate is this performed? (Rate)

To determine this, simply refer to the duration-to-rate conversion chart at the end of this chapter.

Task		Duration (min)	Rate (units/h)
T1	Completed pallet	5	12
T2	Trip to shrink wrapper	3	20
T3	Shrink wrapping a pallet	4	15
T4	Trip to loading area	3	20

1.7 How long each day does the worker perform each task? (Exposure)

To determine this you must first figure out how many times this cycle of tasks is performed.

Task cycle: A series of tasks that is repeated in a sequence.

The entire task cycle takes 15 minutes to complete. By dividing the duty exposure for loading boxes by the task cycle, you arrive at the number of task cycles performed each day.

$$\frac{\text{Duty exposure} = 5\text{ h}}{\text{Task cycle duration} = 0.25\text{ h}}$$

This task cycle is performed 20 times per day. Therefore each task is performed 20 times per day. To determine the exposure for each of the tasks, multiply the time to perform by 20.

Task	Time (Min)	× 20	Exposure Min		Exposure H
T1 Completed pallet	5		100	=	1.6
T2 Trip to shrink wrapper	3		60	=	1.0
T3 Shrink wrapping a pallet	4		80	=	1.3
T4 Trip to loading area	3		60	=	1.0

1.8 How many times does the worker perform this task each day? (Quantity)

To answer this question, multiply the exposure calculated in question 4 by the rate determined in question 3.

Task	Rate × (Units/H)	Exposure =	Quantity (Units/Day)
T1 Completed pallet	12	1.6	19.2
T2 Trip to shrink wrapper	20	1.0	20
T3 Shrink wrapping a pallet	15	1.3	19.5
T4 Trip to loading area	20	1.0	20

This was obviously the long way around. Once you calculated the number of task cycles per day, you already knew the quantity.

Bonus Question

Given that each pallet is made up of 20 boxes, what is the rate and quantity of T1 in boxes per minute?

Bonus Answer

To determine the rate, take the time to perform for an entire pallet load and divide by 20. Then convert using the repetition tables.

5 min/20 = .25.

Using the conversion tables:

0.25 is converted to 4 boxes per minute.

Or simply divide and reduce 20 boxes per 5 minutes to 4 boxes per minute.

20 boxes/5 minutes = 4 boxes per minute.

For the total quantity, simply multiply the number of boxes per pallet (20), by the quantity of pallets per day (20). The answer is 400 boxes per day.

DUTY - TASK QUANTITY

Job Title: *Material handler*
Date: *8/26/87*
Analyst: *Brian Sheehan*

Duty # Task #	Unit of Completion	Time to Perform	Expo	Rate	Quant
D1:T1	*Completed pallet*	*5 min*	*1.6 h*	*12/h*	20
D1:T2	*Shrink trip*	*3 min*	*1.0 h*	*20/h*	20
D1:T3	*Wrapped pallet*	*4 min*	*1.3 h*	*15/h*	20
D1:T4	*Return trip*	*3 min*	*1.0 h*	*20/h*	20

THE BURKE PROTOCOL

EXERCISE 2

Mary Jones assembles toy cars. She places rubber bumpers onto plastic car bodies. This task is performed for 7.25 h each day. The cars are spaced 1 ft apart on the conveyor belt, which moves from left to right in front of her. Mary must assemble every other car, as there is someone else assembling the remaining cars. The belt moves at an average of 9 ft/min.

Questions

Considering the task is car assembly:

2.1 What is the unit of completion for this task?
2.2 What is the time to perform?
2.3 What is the rate ?
2.4 What is the exposure?
2.5 What is the quantity?

Answers

2.1 Unit of completion: 1 toy car
2.2 Time to perform: Approximately 13–14 seconds
2.3 In this case, it is easier to determine the rate first and then consult the conversion table.

Rate: 270 cars/h

Every other car at 9 ft/min means that the rate is 4.5 cars/min. By multiplying 4.5 cars/min by 60 you will get 270 cars/h.

2.4 Exposure: 7.25 h

This is all this person does.

2.5 Quantity: 1958 cars/day

270 cars/h multiplied by 7.25 h/day gives 1958 cars/day.

These exercises should help in understanding the process. The relevancy of this process will become clearer as the risk identification takes place. In the second exercise, if there is a risk factor that occurs only once per task, it will be repeated close to 2000 times per day.

DURATION TO RATE CONVERSION CHART

DURATION SECONDS	RATE REPS/MIN	RATE REPS/HOUR	DURATION SECONDS	RATE REPS/MIN	RATE REPS/HOUR
1	60.00	3,600	31	1.94	116
2	30.00	1,800	32	1.88	113
3	20.00	1,200	33	1.82	109
4	15.00	900	34	1.76	106
5	12.00	720	35	1.71	103
6	10.00	600	36	1.67	100
7	8.57	514	37	1.62	97
8	7.50	450	38	1.58	95
9	6.67	400	39	1.54	92
10	6.00	360	40	1.50	90
11	5.45	327	41	1.46	88
12	5.00	300	42	1.43	86
13	4.62	277	43	1.40	84
14	4.29	257	44	1.36	82
15	4.00	240	45	1.33	80
16	3.75	225	46	1.30	78
17	3.53	212	47	1.28	77
18	3.33	200	48	1.25	75
19	3.16	189	49	1.22	73
20	3.00	180	50	1.20	72
21	2.86	171	51	1.18	71
22	2.73	164	52	1.15	69
23	2.61	157	53	1.13	68
24	2.50	150	54	1.11	67
25	2.40	144	55	1.09	65
26	2.31	138	56	1.07	64
27	2.22	133	57	1.05	63
28	2.14	129	58	1.03	62
29	2.07	124	59	1.02	61
30	2.00	120	60	1.00	60

The Burke Protocol

dur-rep1

DURATION TO RATE CONVERSION CHART

DURATION MINUTES	RATE REPS/HOUR	DURATION MINUTES	RATE REPS/HOUR
1.00	60.00	13	4.62
1.25	48.00	14	4.29
1.50	40.00	15	4.00
1.75	34.29	16	3.75
2.00	30.00	17	3.53
2.25	26.67	18	3.33
2.50	24.00	19	3.16
2.75	21.82	20	3.00
3.00	20.00	21	2.86
3.25	18.46	22	2.73
3.50	17.14	23	2.61
3.75	16.00	24	2.50
4.00	15.00	25	2.40
4.25	14.12	26	2.31
4.50	13.33	27	2.22
4.75	12.63	28	2.14
5.00	12.00	29	2.07
6.00	10.00	30	2.00
7.00	8.57	35	1.71
8.00	7.50	40	1.50
9.00	6.67	45	1.33
10.00	6.00	50	1.20
11.00	5.45	55	1.09
12.00	5.00	60	1.00

The Burke Protocol DUR-REP2

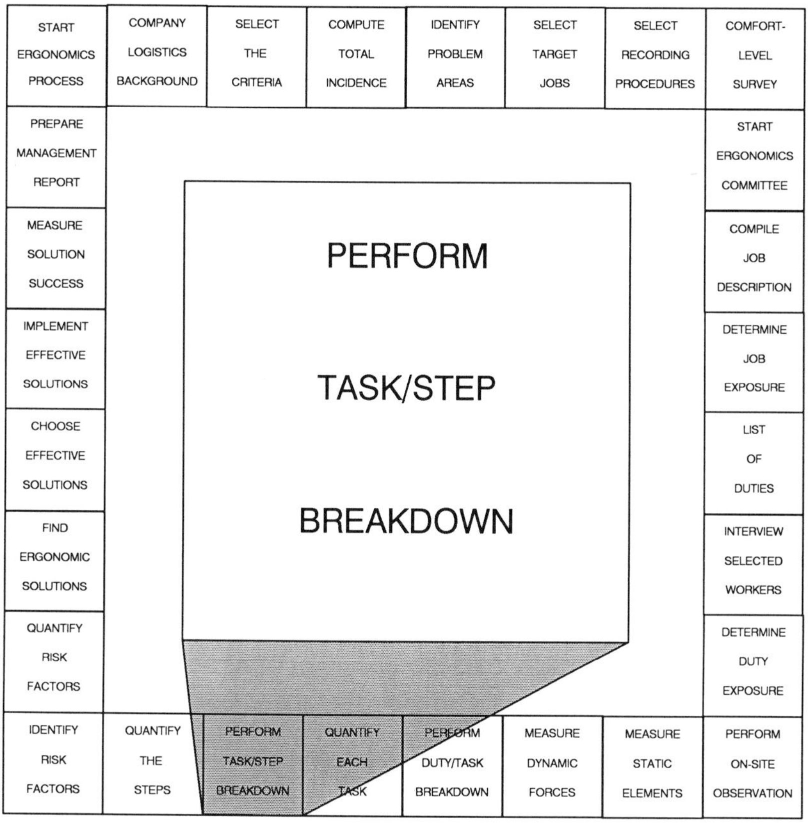

START ERGONOMICS PROCESS

COMPANY LOGISTICS BACKGROUND

SELECT THE CRITERIA

COMPUTE TOTAL INCIDENCE

IDENTIFY PROBLEM AREAS

SELECT TARGET JOBS

SELECT RECORDING PROCEDURES

COMFORT-LEVEL SURVEY

START ERGONOMICS COMMITTEE

COMPILE JOB DESCRIPTION

DETERMINE JOB EXPOSURE

LIST OF DUTIES

INTERVIEW SELECTED WORKERS

DETERMINE DUTY EXPOSURE

PERFORM ON-SITE OBSERVATION

MEASURE STATIC ELEMENTS

MEASURE DYNAMIC FORCES

PERFORM DUTY/TASK BREAKDOWN

QUANTIFY EACH TASK

PERFORM TASK/STEP BREAKDOWN

QUANTIFY THE STEPS

IDENTIFY RISK FACTORS

QUANTIFY RISK FACTORS

FIND ERGONOMIC SOLUTIONS

CHOOSE EFFECTIVE SOLUTIONS

IMPLEMENT EFFECTIVE SOLUTIONS

MEASURE SOLUTION SUCCESS

PREPARE MANAGEMENT REPORT

PERFORM TASK/STEP BREAKDOWN

Perform Task/Step Breakdown

OBJECTIVE

To break a task down into steps and determine the level of exposure.

SUPPLIES

1. Videotape player with pause feature.

GUIDELINES

In some cases it may be appropriate to break each task down into its steps. These steps are also known as physical demands. The physical demands have been defined by the dictionary of occupational titles. *In most cases this is not necessary.* The instances where this may be appropriate would be if the worker performs a single task for all or most of his time. Assembly-line operators, inspectors, or packers may fall into this category. The last example in the previous chapter is such an example.

This is going to be a rather time-consuming and demanding task. It is difficult under ideal conditions and virtually impossible if there are distractions or interruptions. It is extremely helpful if you can operate the videotape player very slowly.

PROCEDURES

Start at the beginning of the task cycle. First focus on the task itself. What is being done? Then, observe each body part as it goes through the activity. Keep referring to the list of physical demands and make a note of when any of them occur. The physical demands are:

Standing: Remaining on one's feet in an upright position without moving about.

Walking: Moving about on foot.

Sitting: Remaining in the normal seated position.

Lifting: Raising or lowering an object from one level to another.

Carrying: Transporting an object, usually holding it in the hands, arms, or shoulders.

Pushing: Exerting a force upon an object so that the object moves away from the force.

Pulling: Exerting a force upon an object so that the object moves towards the direction of the force.

Balancing: Maintaining body equilibrium to prevent falling when walking, standing, crouching, or running on narrow, slippery, or erratically moving surfaces.

Climbing: Ascending and descending ladders, stairs, scaffolding, ramps, poles, and the like using the legs, hands, or arms.

Stooping: Bending body downward and forward by bending spine at waist.

Kneeling: Bending legs at knees to come to rest on knee or knees.

Crouching: Bending body downward or forward by bending legs and spine.

Crawling: Moving about on hands and knees, or hands and feet.

Reaching: Extending the hand(s) and or arm(s) in any direction.

Handing: Seizing, holding, grasping, turning, or otherwise working with hands.

Following is an example of a step breakdown.

Mary Jones assembles toy cars. She places rubber bumpers onto plastic car bodies. This task is performed for 7.25 h each day. The cars are spaced 1 ft apart on the conveyor belt, which moves from the left to right in front of her. Mary must assemble every other car, as there is someone else assembling the remaining cars. The belt moves at an average of 9 ft/min.

List of Duties

D1 Assemble toy cars

Duty/Task Breakdown

D1 Assemble toy cars
T1 Assemble toy cars

Task Step Breakdown

Task 1: Assemble Cars

Step 1 Reaches for car with left hand
Step 2 Reaches into parts bin for bumper with right hand
Step 3 Handles the pieces together to assemble
Step 4 Reaches back to the belt to replace car

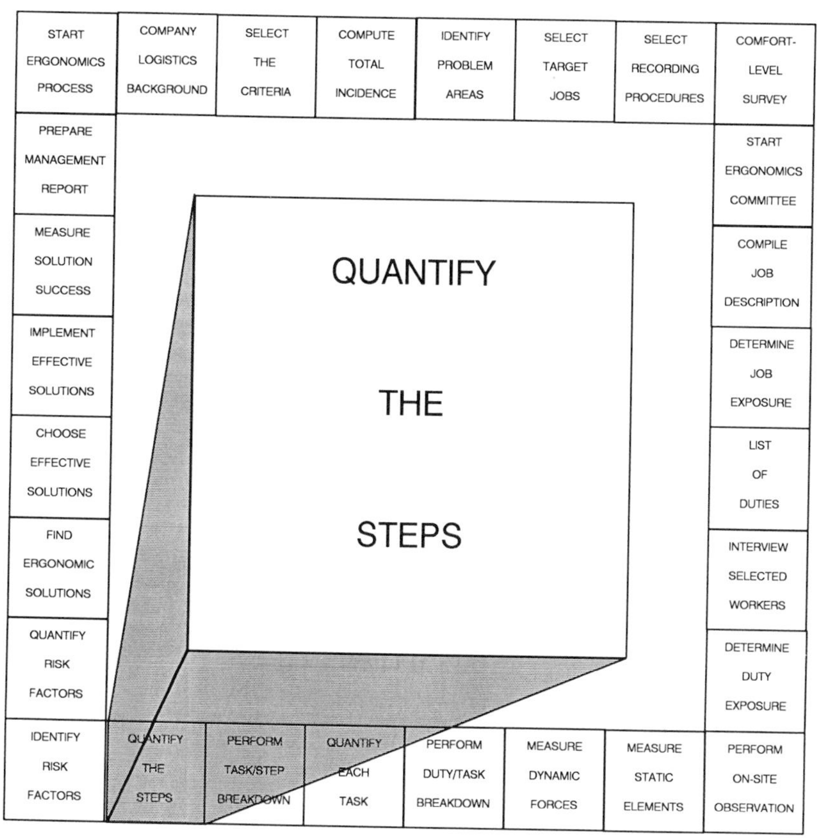
START ERGONOMICS PROCESS
COMPANY LOGISTICS BACKGROUND
SELECT THE CRITERIA
COMPUTE TOTAL INCIDENCE
IDENTIFY PROBLEM AREAS
SELECT TARGET JOBS
SELECT RECORDING PROCEDURES
COMFORT-LEVEL SURVEY
START ERGONOMICS COMMITTEE
COMPILE JOB DESCRIPTION
DETERMINE JOB EXPOSURE
LIST OF DUTIES
INTERVIEW SELECTED WORKERS
DETERMINE DUTY EXPOSURE
PERFORM ON-SITE OBSERVATION
MEASURE STATIC ELEMENTS
MEASURE DYNAMIC FORCES
PERFORM DUTY/TASK BREAKDOWN
QUANTIFY EACH TASK
PERFORM TASK/STEP BREAKDOWN
QUANTIFY THE STEPS
IDENTIFY RISK FACTORS
QUANTIFY RISK FACTORS
FIND ERGONOMIC SOLUTIONS
CHOOSE EFFECTIVE SOLUTIONS
IMPLEMENT EFFECTIVE SOLUTIONS
MEASURE SOLUTION SUCCESS
PREPARE MANAGEMENT REPORT
QUANTIFY
THE
STEPS

Quantify the Steps

OBJECTIVE

To determine the level of exposure for each step as expressed in time (duration), frequency (rate), and total number or repetitions per day (quantity).

This step is for the most part identical to the process described in the section entitled Quantify Each Task.

PROCESS

Define a unit of completion for the step being analyzed, determine how long it takes to perform that unit of completion, verify how long this step is performed in terms of time, and verify how many times this step is performed.

GUIDELINES

This rather intensive type of analysis will likely be reserved for jobs which have a very limited number of duties or tasks, or where a single task or task cycle takes up almost all the job exposure.

The need to quantify the steps will also depend on what is to be done with the information. In most cases, the relationship between the steps and the task is clear enough to be able to determine the rate, quantity, and exposure. For example, a step may be performed three times a task. With this information, you can refer to the task quantities and use that relationship to quantify each step.

You also may choose to sort out the various physical demands and add up the exposures or repetitions. This can be helpful in determining the physical demands for a job description. A worker may "lift boxes" to unload a truck in the morning for two hours, and then "lift boxes" for an hour as a part of putting stock away in the afternoon. This would allow you to state that a physical demand of this job is the ability to lift boxes for a total of three hours a day.

The process as described here is not intended and may not be sufficient to determine job demands for the purpose of setting employment criteria. It is suggested that you seek appropriate legal counsel if that is your goal.

These procedures will work well for the purpose of an analysis where the goal is to determine less stressful processes by identifying ergonomic risk factors.

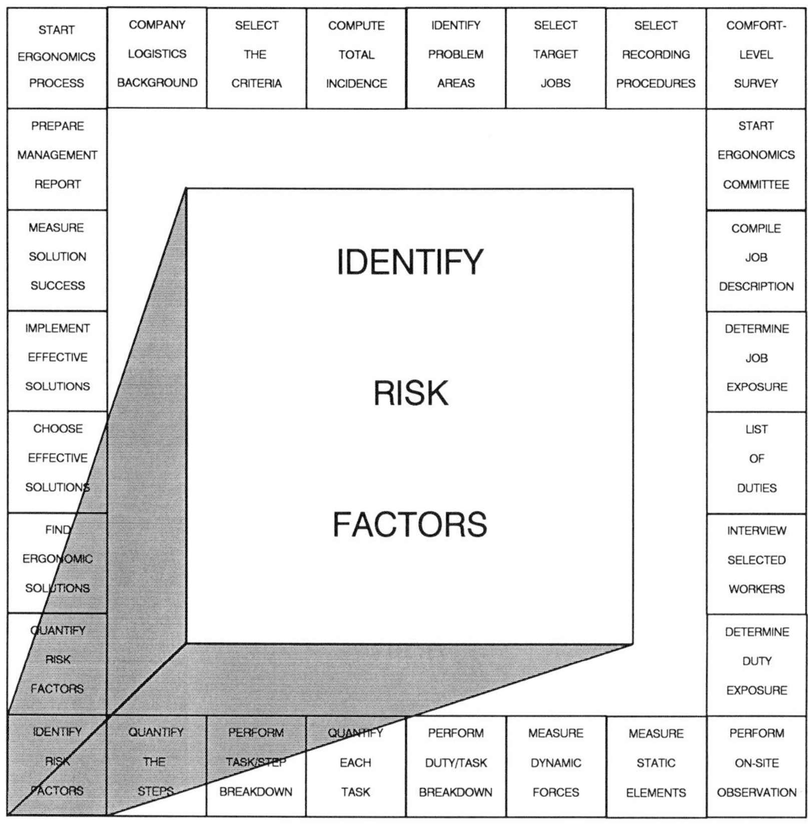

START ERGONOMICS PROCESS
COMPANY LOGISTICS BACKGROUND
SELECT THE CRITERIA
COMPUTE TOTAL INCIDENCE
IDENTIFY PROBLEM AREAS
SELECT TARGET JOBS
SELECT RECORDING PROCEDURES
COMFORT-LEVEL SURVEY
START ERGONOMICS COMMITTEE
COMPILE JOB DESCRIPTION
DETERMINE JOB EXPOSURE
LIST OF DUTIES
INTERVIEW SELECTED WORKERS
DETERMINE DUTY EXPOSURE
PERFORM ON-SITE OBSERVATION
MEASURE STATIC ELEMENTS
MEASURE DYNAMIC FORCES
PERFORM DUTY/TASK BREAKDOWN
QUANTIFY EACH TASK
PERFORM TASK/STEP BREAKDOWN
QUANTIFY THE STEPS
IDENTIFY RISK FACTORS
QUANTIFY RISK FACTORS
FIND ERGONOMIC SOLUTIONS
CHOOSE EFFECTIVE SOLUTIONS
IMPLEMENT EFFECTIVE SOLUTIONS
MEASURE SOLUTION SUCCESS
PREPARE MANAGEMENT REPORT
IDENTIFY RISK FACTORS

Identify Risk Factors

OBJECTIVE

To compose a list of ergonomic risk factors associated with specific duties, tasks, or steps.

SUPPLIES

1. Videotape player with slow and/or pause feature.
2. Videotape of operation to be analyzed.
3. Risk identification guide.

GUIDELINES

The goal of this step is to compile a list of risk factors that have been associated with a specific duty, task, or step. Ergonomic risk factors are conditions that are likely to act as barriers to optimal productivity and consistent quality, and have been associated with a high incidence of cumulative trauma disorders. The overall goal of the ergonomics process is to eliminate as many unnecessary risk factors as possible.

The approach to risk identification presented here will focus on the recognition of conditions or states that vary from an ideal. Quantifying each risk factor will be limited to forces, repetitions, and exposures. This approach will be especially helpful to persons performing this type of analysis with limited time. It will allow them to focus on developing strategies and arriving at solutions to correct the conditions.

You may choose to spend more time measuring the degree of each of the risk factors and still utilize this approach. The decision to measure and quantify each risk factor should be based on your personal philosophy and experience. Some may feel that you don't need to weigh an item to know that you can decrease the stress by decreasing the weight of the item. Others may argue that you need a computer-generated model complete with digitized video and electromyographic input to assess the level of the stress prior to considering various interventions.

The following is meant only as a very brief overview of ergonomic risk factors. Extensive material is available supporting that these conditions are causative in the development of several cumulative trauma disorders. Specific references can be found in the bibliography in the appendix.

A *cumulative trauma disorder* is a breakdown of a body part as a result of repetitive overuse, overexertion, or excessive stress. Examples of conditions that can be referred to as cumulative trauma disorders are:

Bursitis
Carpal tunnel syndrome
Cubital tunnel syndrome
DeQuervain's disease
Digital neuritis
Epicondylitis
Excessive fatigue
Frozen shoulder
Gamekeeper's thumb
Ganglion cyst
Guyon canal syndrome
Hypothenar hammer syndrome
Peritendonitis crepitans
Pronator teres syndrome
Radial tunnel syndrome
Raynaud's phenomena
Stenosing tenosynovitis
Stenosing tenosynovitis crepitans
Synovitis
Supinator syndrome
Tendonitis
Tenosynovitis
Thoracic outlet syndrome

In the simplest terms, cumulative trauma disorders develop as a result of long-term exposure to certain activities or environmental conditions that leads to an adaptive response within the body. Currently there is no scientific data that define "long term." The evidence that certain working or environmental conditions are risk factors for the development of cumulative trauma disorders is the result of retrospective studies. The purpose of an ergonomics analysis is to identify and limit the risk-factor exposure level to the safest possible for all workers.

Just about all risk factors can be placed into one of six categories. They are:

1. End range position
2. Unsupported posture
3. Forceful exertion
4. Environmental factors
5. Excessive metabolic (energy) demand
6. Repetition (?)

End Range Position

End range position refers to moving a joint in the body as far as it will go or close to its maximum. A waiter who carries a tray of food overhead through a crowded restaurant will bend his wrist back into an end range position. Reaching overhead at the shoulder to retrieve or replace an object on a high shelf would be an end range position at the shoulder. Looking down at a desk or table all day would be an end range position for the neck. The stress here is on the muscles, tendons, ligaments, and other structures within the body as they are stretched to their "end range."

The end range position is measured by the amount of available range of motion at the joint being used. For the most part, movement in the last 10% (end range) of motion would be considered a serious deviation. The optimal would be any movement in 5–10% of the body's naturally strongest and least stressful position.

There is a great deal of variation from one person to another with regard to end range. In addition, each individuals range of motion varies depending on the side of the body. Therefore, it is not practical to list acceptable ranges for the various joints in the body.

End range positions can be tracked and identified by attaching a device to a person's body to measure and record his or her movement, or through direct observation. Direct observation is much more practical and can be facilitated by using the risk identification guide at the back of this chapter.

Sustained or Unsupported Posture

Any time a worker must maintain a position for an extended period of time, this is considered to be a sustained posture. If a person must forward bend at the waist even a little to work at his or her workstation and must maintain that position while he or she works, it is considered as unsupported posture. If a person must hold his or her arm out to his or her side or in front, that would also be considered an an unsupported posture. If a person must hold an object in place, perhaps while he or she works on it with his or her other hand, that too is considered an unsupported sustained posture.

The only way to identify this risk factor is through direct observation. This identification can be facilitated by using the risk identification guide at the end of this chapter.

Forceful Exertion

Forceful exertion deals with overcoming an outside agent to perform some activity. In most cases, that outside agent is gravity. If a worker is required to lift a box from the floor to a waist-high table, he must overcome a gravitation force equal to the weight of the box. There are tasks in which gravity does not play a role in creating the force that needs to be overcome. An example would be having to forcefully remove or insert small rods from tightly fitted machine parts.

Measuring and quantifying the forces can be simple or very complex. There are a couple of options regarding how to quantify these forces. The simplest method is to weigh the item being handled. The disadvantage is that it does not take into account muscular force lever or biomechanical considerations.

Variables such as the method of handling, the size of the box, and hand coupling will result in several different forces to be overcome at various points in the body to perform this task. You might utilize a series of formulae to resolve force vectors and to determine the torques and forces associated at this joint. This will provide you with accurate information but can be very time consuming. Another approach is to use the NIOSH formula (see Appendix E), which allows you to determine in general terms certain limits for lifting, but does not address any other activities.

There are also several computer models available to arrive at pure empirical data. These allow you to gather an enormous amount of data relative to the forces at various parts of the body during certain lifting, pushing, or pulling activities. This approach is valuable from a research standpoint; however, the positions of the body and activities that can be analyzed are limited.

Methods for the measurement of forces using a more universal and practical method were discussed in the section entitled Measure Dynamic Forces. Identification of these forces is simplified in the risk identification guide.

Environmental Factors

Environmental factors take into consideration a rather extensive combination of variables. You are concerned here with the way the body interacts with its surroundings. Examples of environmental factors are cold, heat, moisture, noise, vibration, and man-machine coupling. All are self-explanatory, except man-machine coupling. Included in this area are such concerns as sharp edges of tables or tools that do not fit the operator.

Examples of various environmental risk factors and where they are likely to be found are listed below.

Heat: Foundry workers typically work in very hot environments and handle hot materials.

Cold: Butchers may handle cold materials in a very cold storage facility.

Moisture: Dishwashers work in water medium in typically humid kitchen areas.

Man-machine coupling: Manufacturing assemblers work with straight-line tools at a sharp-edged table on a very hard surface.

Vibration: Auto mechanics are exposed to segmental vibration when operating a hand-held impact wrench; a truck driver is exposed to whole-body vibration when driving cross-country.

The measure of the environmental factors will be determined by the particular variables involved. To keep this process practical, it is best to focus more on the identification and less on the measurement. In some cases, the utilization of environmental engineers and industrial hygienists will be necessary.

The risk identification guide facilitates the recognition of these potentially hazardous ergonomic conditions.

Excessive Metabolic/Energy Demand

Certain jobs place an excessive physiological demand for energy upon the body, either for short, highly demanding periods or for extensive, moderately demanding periods. The degree of this particular risk factor will vary with the individual worker, the level of physiological demand, and the time period that it must be maintained.

A person's ability to perform work is measured by the amount of energy he or she can produce. The term *work,* as it is used here, means any type of physical exertion. Each person has a maximum amount of energy he or she can produce to perform work. Levels of physical exertion for each person can be related to the percentage of maximum energy production.

A person can safely work for 8 h at about 33% of his maximum energy output.

Identification of excessive energy requirements can be performed utilizing physiological monitoring or computer equipment. You can likewise consult some of the many charts and formulae that have been developed to assist in predicting the energy requirements of various tasks. A more practical approach is to identify certain behaviors that are indications of fatigue. Excessive sweating and the inability to talk in a normal manner are examples of signs of excessive physical demands. These behaviors are outlined at the end of the risk identification guide.

Repetition

Repetition is any action or activity that is repeated. Rather than being considered a risk factor, it is actually a quantifier of the other risk factors. If an individual must forward bend at the waist to pick up a 75-pound box one time, that is considered stressful. If that individual must perform that task repetitively, then the number of repetitions is a factor of that stress. Put simply, if you have to do something stressful once an hour, it may be bad. If you have to perform that repetitively for several hours or days, that's much worse.

The resolution of the problem of repetitious exposure to one of the above risk factors is also rather straightforward. Either reduce the number of repetitions or reduce the risk factor itself.

PROCEDURES FOR THE IDENTIFICATION OF ERGONOMIC RISK FACTORS

There are several approaches to identifying risk factors. You can use a checklist. This can be performed quickly and will identify a limited number of specific conditions. The disadvantage is that in order to facilitate the quick preference, the list must be relatively short. By maintaining a short list, it is possible to overlook risk factors that are not on the list.

Another approach is to base everything on certain established standards. For instance, a table should be a certain number of inches high if a worker is performing paperwork. If the worker is performing heavy assembly work, then the standard says that it should be a different height. These standards are based on size norms for a range of persons and therefore do not take into consideration a unique worker population.

A purely biomechanical approach would be to examine the stresses and forces on each joint and body part. This approach limits itself to forceful exertion and even postural stresses, potentially missing other important factors.

To assist you in developing the observation and analysis skills to perform a comprehensive ergonomic analysis, an integrated approach is most beneficial. The following procedures are designed to help you feel comfortable with this integrated approach. The risk identification guide will help you through the process of developing the most complete list of ergonomic risk factors. Individual sections of the worker's body are highlighted for analysis. Then the presence of various types of risk factors are screened.

For example, you might start by looking at the worker's foot and ankle. Record any end range positions, unsupported postures, forceful exertions, or other environmental risk factors. As each risk factor is identified, record it and any necessary additional information that will help in quantifying the factor. This should be done with a completely open mind. Do not limit yourself by an initial perception of the degree or severity of the risk factor or the likelihood of modification. Even though you may have prior knowledge or experience that says that there's nothing you can do about a particular risk factor, it should still be documented. At this point you don't want to limit your perspective in any way.

To quantify the risk factors you will take advantage of the duty and task breakdown preparation steps. Once you have identified risk factors associated with a particular duty, task, or step, it is reasonably simple to establish the relationship between the risk factor and the duty, task, or step. For example, if a painter must reach overhead to an end range position of the shoulder twice during the performance of a task, then the number of repetitions is simply the **quantity** for that task multiplied by 2.

Additional information for quantifying this step would be the weight of the brush or roller that the painter is using. Assuming you have already gathered all the dynamic measurements, once you recognize a risk factor, simply refer to the previous work and all the quantifying information should be there.

With these basics in mind, watch a worker performing his or her job or watch the videotape of the worker and systematically note the risk factors demonstrated. It is essential to stay focused on one body part or area at a time. Go through the entire process of identifying end range positions, unsupported sustained postures, forceful exertions, and environmental factors. Make any additional notes and then move on to the next body part. When you have covered all the areas of the body, focus on the aspect of excessive metabolic or energy demands.

There is usually a great temptation to try to come up with suggested interventions to solve or reduce each risk factor as you proceed. You may choose to make a note off to the side about an intervention, but it is more important to stay focused on a single body part and the associated stresses.

To facilitate the performance of the risk identification, use the risk identification guide at the end of this chapter. Some of the information in this chapter will be repeated. This guide is meant to be a reference for on-going use and review. This can be helpful if you perform ergonomic analysis on an infrequent basis.

Whether you choose to use the procedures outlined in the risk identification guide or not, the following format can be useful in writing up an interim report.

RISK IDENTIFICATION — NARRATIVE FORMAT

(Company Name)_______________
(Job Title)____________________ __________________ (Analyst Name)
(Job Number)__________________ __________ (Date)

While performing the task of ________(task)________,
which is performed ___(task expo)___ h/d, at a rate of ___(task rate)___
/h (min) the subject must

Risk Factors

1. ....
2. ....
3. ....
4. ....

RISK IDENTIFICATION — NARRATIVE FORMAT

AAA Inc.
Order Picker
006-34

Mike Burke
3-15-91

While performing the task of getting orders from stock,
which is performed 4 h/day,
at a rate of 20 orders/h,
for a total of 80 orders/day,
the subject is exposed to the following risk factors:

1. Reaching overhead at shoulder repetitively
2. Prolonged exposure to cement floor at foot
3. Forward bending at waist to lift 32-lb. boxes repetitively
4. Holding wrist in end range position to the side

These risk factors have been associated with an increased incidence of cumulative trauma disorders.

ERGONOMIC RISK IDENTIFICATION GUIDE

RISK IDENTIFICATION GUIDE

Principles

Ergonomic risk factors are conditions that are likely to act as barriers to optimal productivity and consistent quality, and have been associated with a high incidence of cumulative trauma disorders.

The ergonomic risk factors are:

1. End Range Position
2. Unsupported Sustained Posture
3. Forceful Exertion
4. Environmental Factors
5. Excessive Metabolic (Energy) Demand

1. End Range Position

This refers to moving a joint in the body as far as it will go or close to it. Examples are backward bending of the wrist, twisting at the waist, wrist flexion, and even big-toe extension. This can be measured either by attaching a device to someone's body to monitor his motions or by direct observation.

2. Sustained Unsupported Posture

Any time a worker must maintain an unsupported position for an extended period of time, it is considered a sustained posture. Any position can be a sustained posture. Examples are working bent over, holding arms out to the side, looking down while working at desk, sitting without a back rest, cutting hair, or even standing all day. The most effective method for measuring this is direct observation.

3. Forceful Exertion

Force deals with the outside agent that must be overcome to perform a step, task, or duty. In most cases, that outside agent is either gravity or friction. Examples are lifting a box, pushing a cart, fitting two pieces of tubing together, pulling on a rope or chain, or screwing in a bolt.

Measurement and analysis can be achieved using computer models, the NIOSH Formula [AL = 90(6/H)(1 - 0.1 | V-30|)(.7 + 3/D)(1 - F / Fmax)], or direct measurement using either a force gauge dynamometer or a scale.

4. Environmental Factors

Environmental factors take into consideration the way the body interacts with its surroundings. Examples are cold, heat, moisture, vibration, and man-machine coupling. Measurement can be through direct observation without measurement or through industrial hygiene and environmental monitoring.

5. Excessive Fatigue

This can be defined as an excessive physiological demand for energy placed on the body, either for short, highly demanding periods or extensive, moderately demanding periods. Measurement can be accomplished through computer models, telemetry, the Borg scale, or direct observation of visual signs.

RISK IDENTIFICATION GUIDE

Procedures

Determine and record the task to be observed on the risk identification form.

Observe the worker either in person or on videotape.

Focus on a specific body part.

Identify any risk factors present FOR THAT BODY PART.

Answer any specific questions as directed by the guide on the risk identification form.

When all questions have been answered for this body part, move on to the next one.

When all body parts have been observed answer question 6.

Move to the next task to be studied, and identify and record the duty and task number.

RISK IDENTIFICATION GUIDE

Questions

A. What task or duty is being performed?

B. What body part is this?

1. Is this an end range position?

   1a. What position or motion is this?

2. Is this body part being maintained in an unsupported position?

   2a. What position or motion is this?

   2b. How much Force(Weight) must be overcome?

3. Is there force being applied?

   3a. Is this body part being held in a static position against force?

      3a1. What position or motion is this?

      3a2. How much Force(Weight) must be overcome?

   3b. Is this a lifting force?

      3b1. How much Force(Weight) must be overcome?

   3c. Is this a pulling force?

      3c1. How much Force(Weight) must be overcome?

   3d. Is this a pushing force?

      3d1. How much Force(Weight) must be overcome?

   3e. Is this a twisting force?

      3e1. How much Force(Weight) must be overcome?

4. Is this body part exposed to environmental hazards?

    4a. Is this body part exposed to Heat?

    4b. Is this body part exposed to Cold?

    4c. Is this body part exposed to Hard Surface?

    4d. Is this body part exposed to Sharp Surface?

    4e. Is this body part exposed to Vibration?

    4f. Is this body part exposed to some other condition?

        4f1. What is the other condition?

5. Is this a highly repetitious motion?

    5a. What position or motion is this?

    5b. What is the rate of the repetitions per task?

RETURN TO QUESTION B

WHEN ALL BODY PARTS ARE COMPLETED GO TO QUESTION 6

6. Is there an indication that this is a high energy demand job?

    6a. Profuse sweating?

    6b. Unable to speak normally?

    6c. Reaching for support?

    6d. Rubbing or massaging a body part?

        6d1. What body part is this?

Job Title	*Order picker*			**RISK FACTOR IDENTIFICATION WORKSHEET**
Date	*4/21/81*			
Analyst	*Bill Wilson*			

Duty:Task *Pulling stock to fill orders*

Expo *4 h/day* Rate *20 orders/h* Quantity *80 orders/day*

R1	*Moves arm overhead at shoulder*					
Expo	*4 h/day*	Rate	*1 / Order*	Quantity	*80 / day*	
R2	*Prolonged exposure to cement floor*					
Expo	*4 h/day*	Rate		Quantity	*4 h/day*	
R3	*Forward bends at waist to pick up 32# box*					
Expo	*4 h/day*	Rate	*2 / Order*	Quantity	*160/day*	
R4	*Holds wrist sidebent to move collection bin*					
Expo	*4 h/day*	Rate	*1 / Order*	Quantity	*80 / day*	
R5						
Expo		Rate		Quantity		
R6						
Expo		Rate		Quantity		
R7						
Expo		Rate		Quantity		
R8						
Expo		Rate		Quantity		
R9						
Expo		Rate		Quantity		

THE BURKE PROTOCOL

Body Part Movement Identification

Body Part	Movement	Term
Ankle	Points toes/pushes up on toes	Plantarflexion
	Pulls end of foot up	Dorsiflexion
	Turns foot in	Inversion
	Turns foot out	Eversion
Knee	Bends knee	Flexion
	Straightens knee	Extension
Hip	Pulls knee up towards body	Flexion
	Moves whole leg back at hip	Extension
	Moves leg out to side	Abduction
	Moves leg across in front	Adduction
	Turn leg in at hip	Internal rotation
	Turn leg out at hip	External rotation
Low back	Bends at waist	Flexion
	Bends backwards at waist	Extension
	Twists at waist	Rotation
	Bends to the side	Sidebend
Neck	Bends neck down or forward	Flexion
	Bend neck up or back	Extension
	Move head to side	Sidebend
	Turns head	Rotation
Shoulder	Moves arm overhead	Flexion
	Moves arm back or behind	Extension
	Moves arm to side & overhead	Abduction
	Moves arm across body	Adduction
	Slumps shoulder forward	Protraction
	Pulls shoulder backward	Retraction
	Turns arm in at shoulder	Internal Rotation
	Turns arm out at shoulder	External Rotation
Elbow	Bends elbow	Flexion
	Bends elbow backward	Extension
Wrist	Bends wrist down	Flexion
	Bends wrist backward	Extension
	Bends wrist (to pinky side)	Ulnar deviation
	Bends wrist (to thumb side)	Radial deviation
	Turns hand over (down or in)	Pronation
	Turns hand over (up or out)	Supination
Hand	Closes hand into fist	Flexion
Fingers	Closes hand into fist	Flexion
	Bends fingers backward	Extension

ANKLE MOVEMENTS

Points toes. (Ankle plantarflexion.)

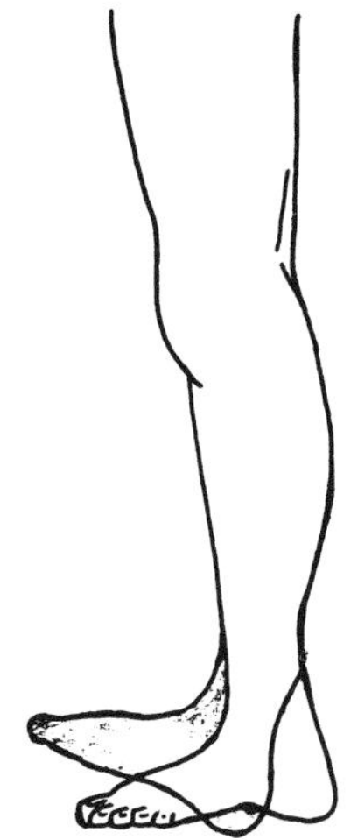

Pulls foot up. (Ankle dorsiflexion.)

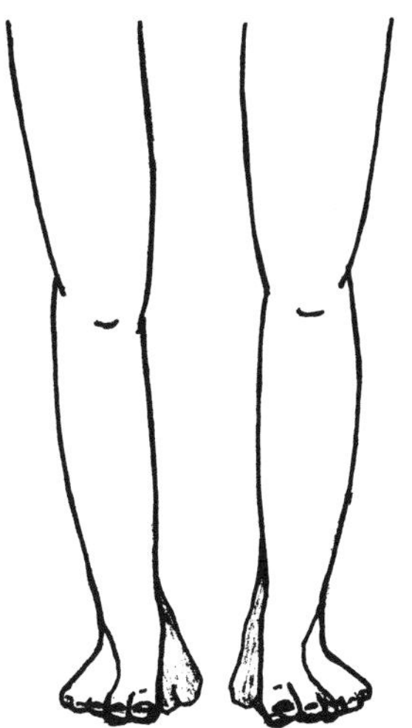

Turns foot in. (Ankle inversion.)

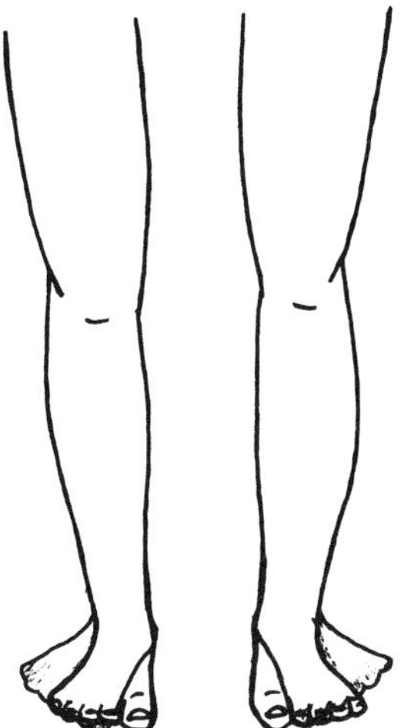

Turns foot out. (Ankle eversion.)

KNEE MOVEMENTS

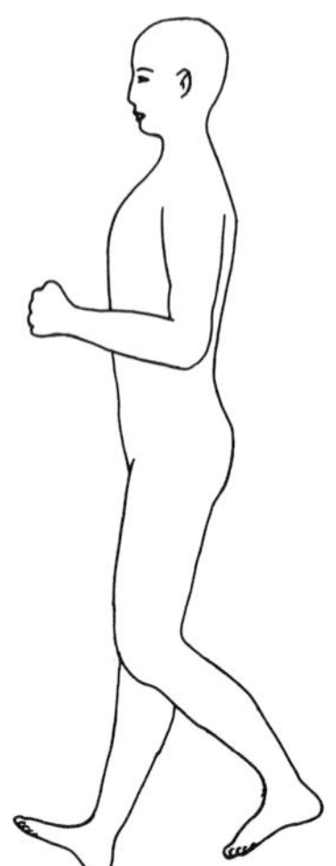

Bends or flexes knee. (Knee flexion.)

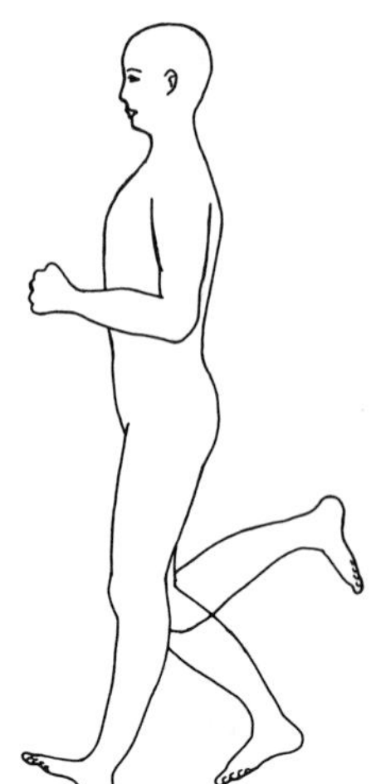

Bends or straightens knee. (Knee extension.)

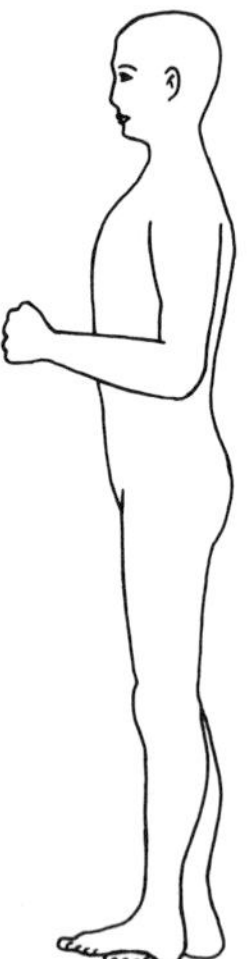

Bends knee backwards. (Knee hyperextension.)

HIP MOVEMENTS

Pulls knee towards body. (Hip flexion.)

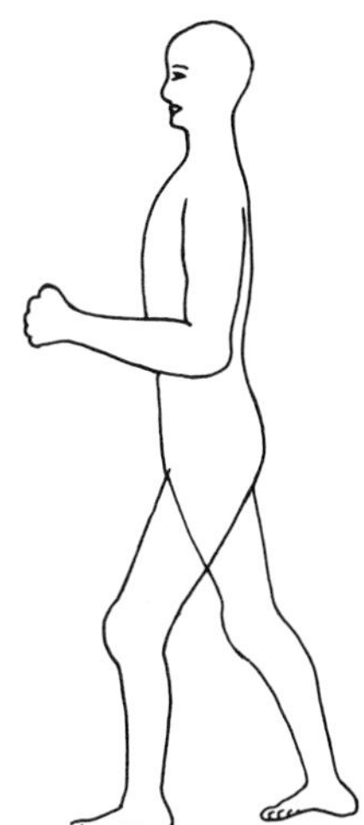

Moves whole leg back at hip. (Hip extension.)

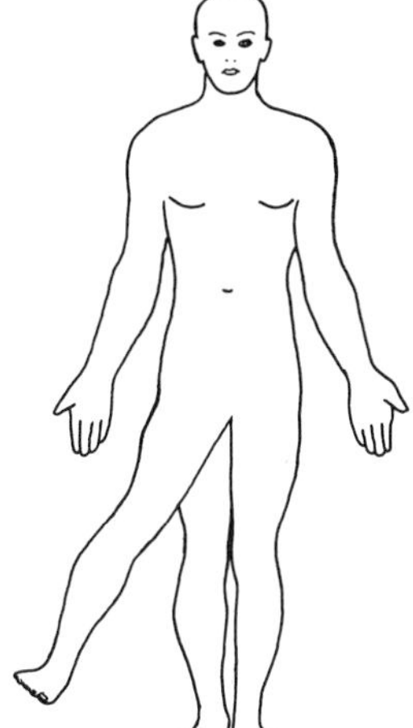

Moves leg out to side. (Hip abduction.)

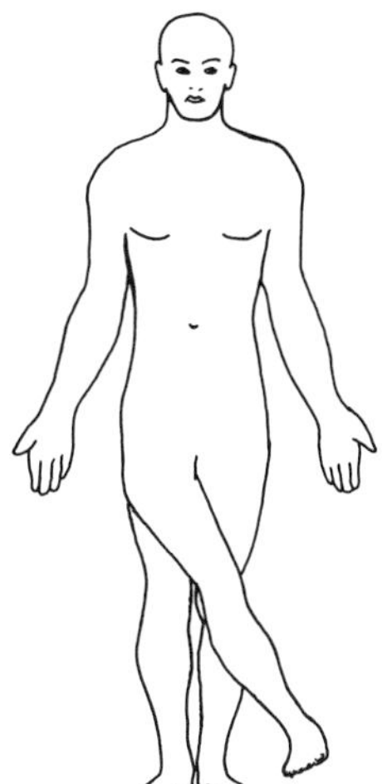

Moves leg across in front. (Hip adduction.)

HIP MOVEMENTS (CONTINUED)

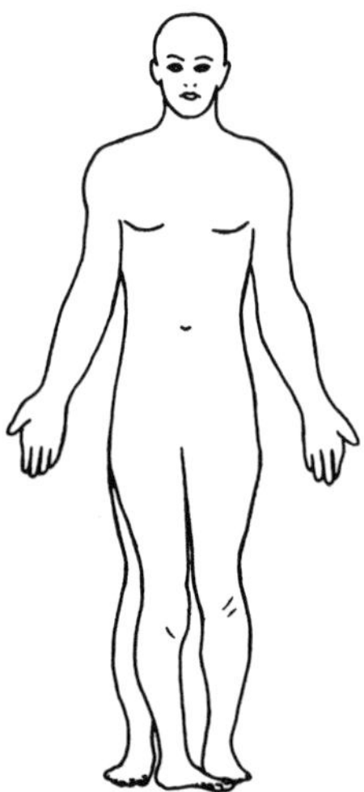

Turns leg in at hip. (Hip internal rotation.)

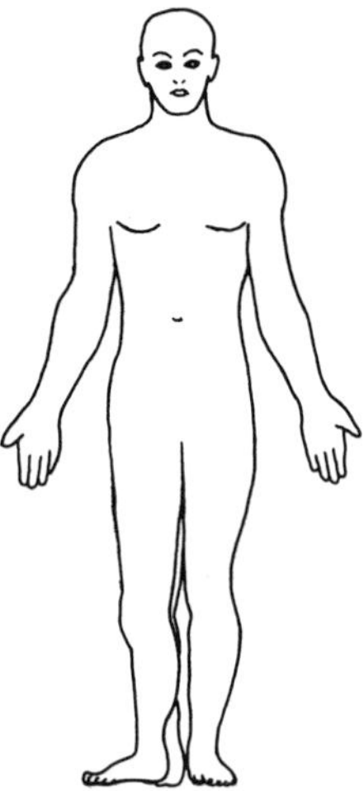

Turns leg out at hip. (Hip external rotation.)

LOW BACK MOVEMENTS

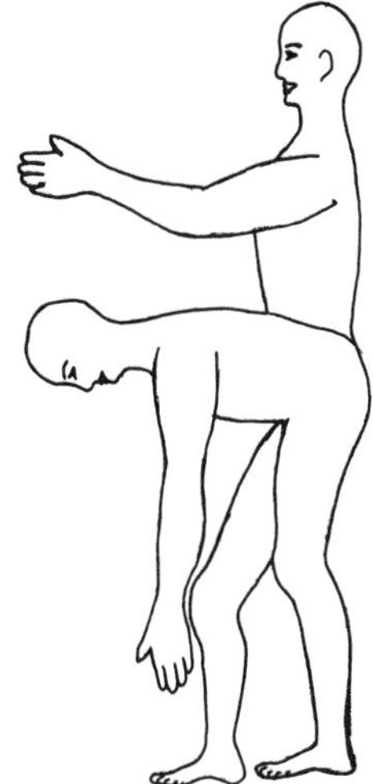

Bends at waist. (Low back flexion.)

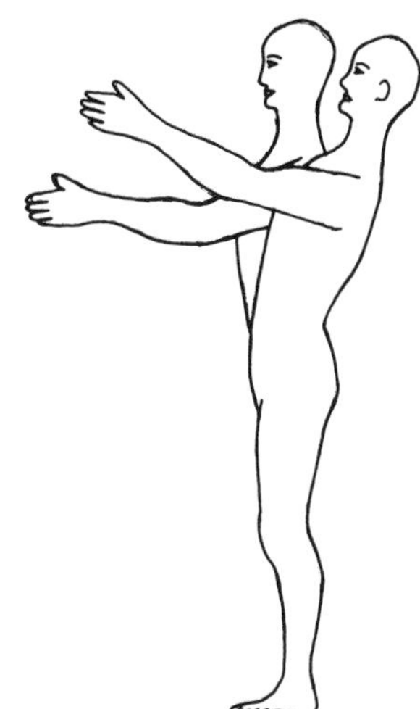

Bends backward at waist. (Low back extension.)

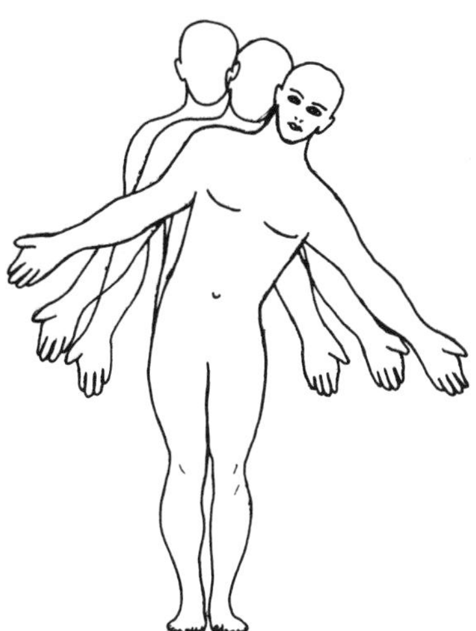

Bends to the side. (Low back sidebending.)

NECK MOVEMENTS

Bends neck down/forward. (Neck flexion.)

Bends neck back/up. (Neck extension.)

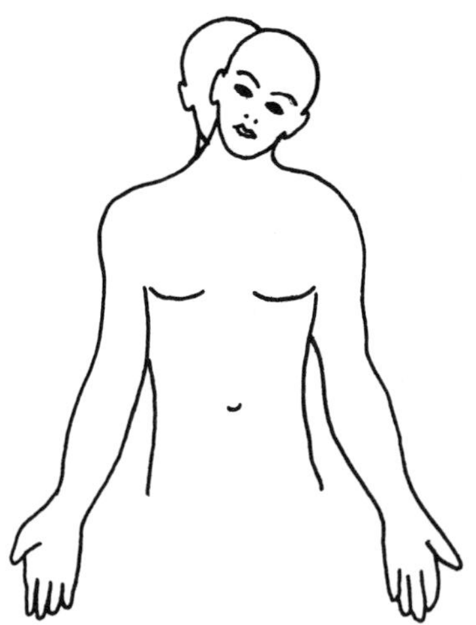

Moves head to side. (Sidebends neck.)

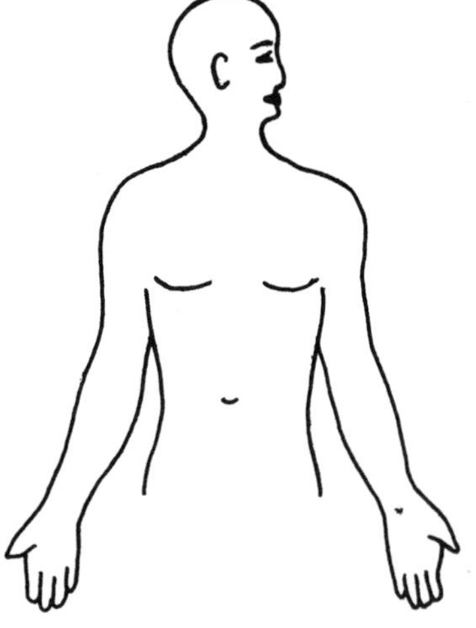

Turns head. (Neck rotation.)

SHOULDER MOVEMENTS

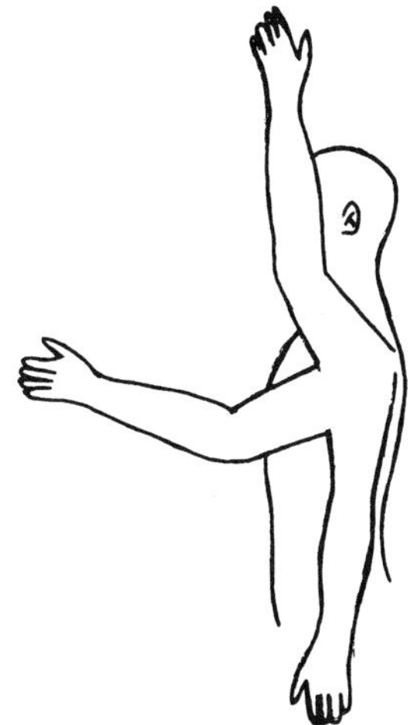

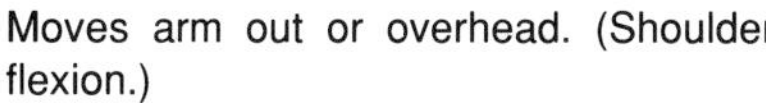

Moves arm out or overhead. (Shoulder flexion.)

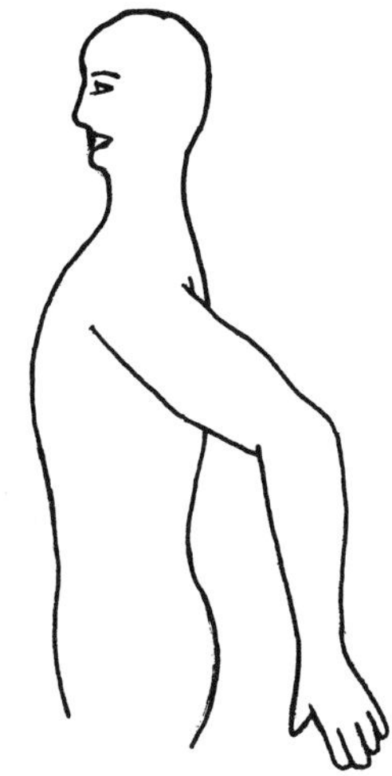

Moves arm back or behind. (Shoulder extension.)

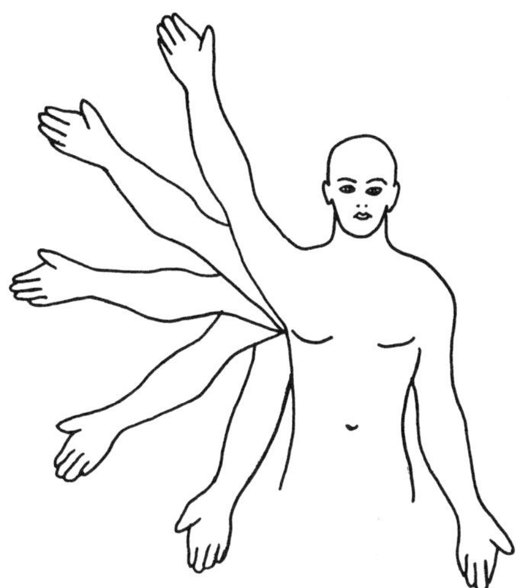

Moves arm out to side/overhead. (Shoulder abduction.)

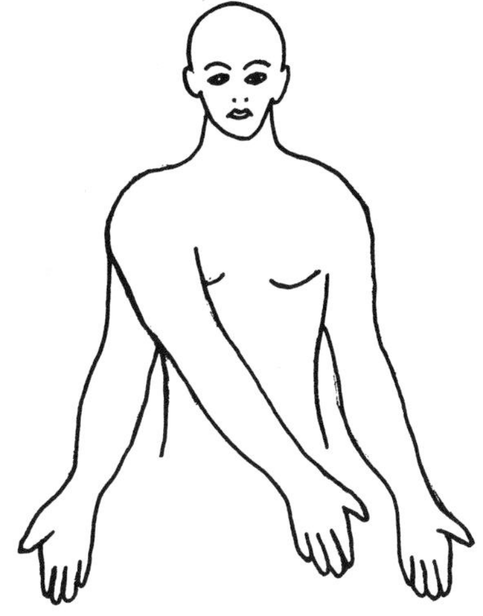

Moves arm across body. (Shoulder adduction.)

SHOULDER MOVEMENTS (CONTINUED)

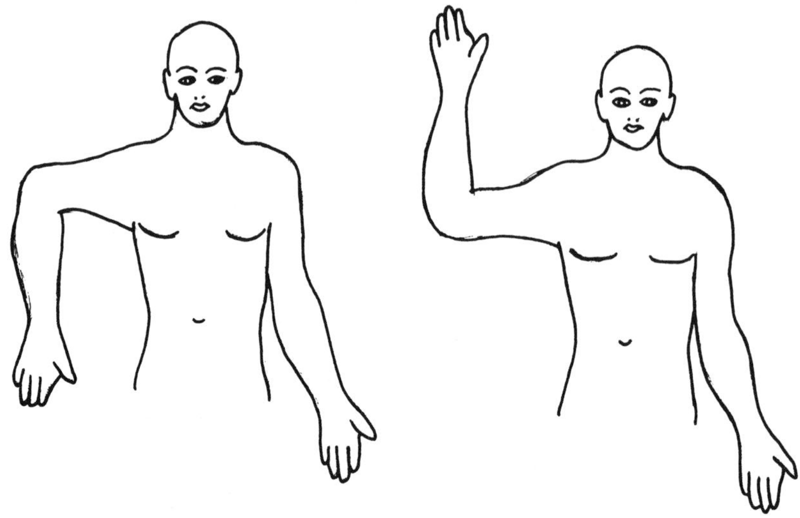

Turns arm in at shoulder. (Shoulder internal rotation.)

Turns arm out at shoulder. (Shoulder external rotation.)

Shrugs/elevates shoulders. (Shoulder elevation.)

ELBOW MOVEMENTS

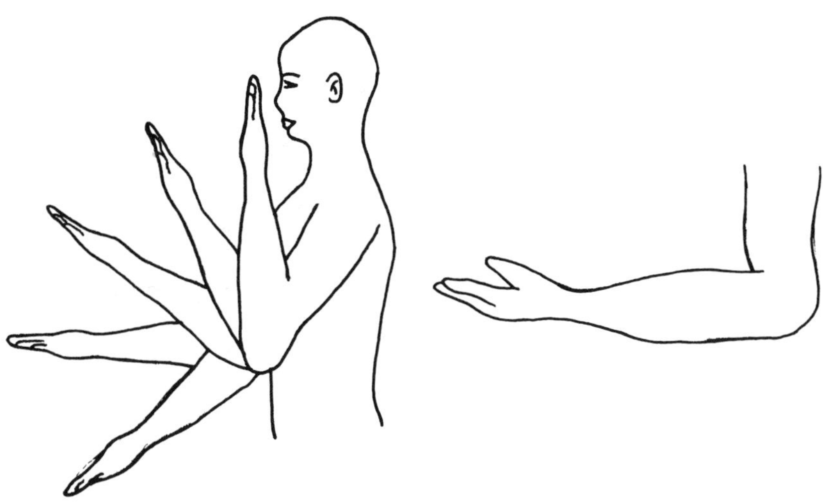

Bends elbow/straightens elbow. (Elbow flexion/extension.)

Turns palm up/back. (Forearm supination.)

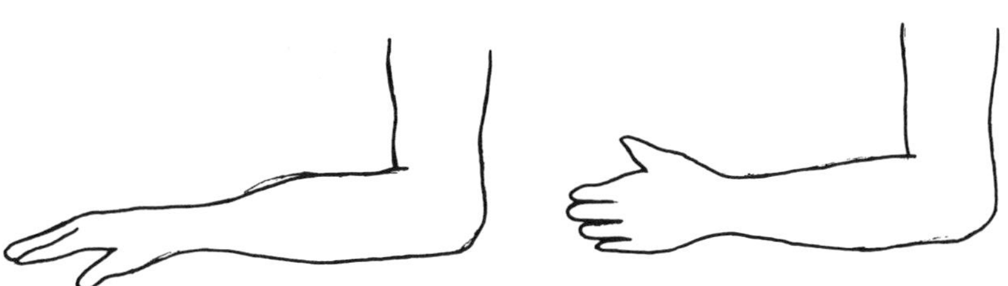

Turns arm down/forward. (Forearm pronation.)

Neutral elbow. (No supination or pronation.)

WRIST MOVEMENTS

Bends wrist down. (Wrist flexion.)

Bends wrist backward. (Wrist extension.)

Bends wrist to pinky side. (Wrist ulnar deviation.)

Bends wrist to thumb side. (Wrist radial deviation.)

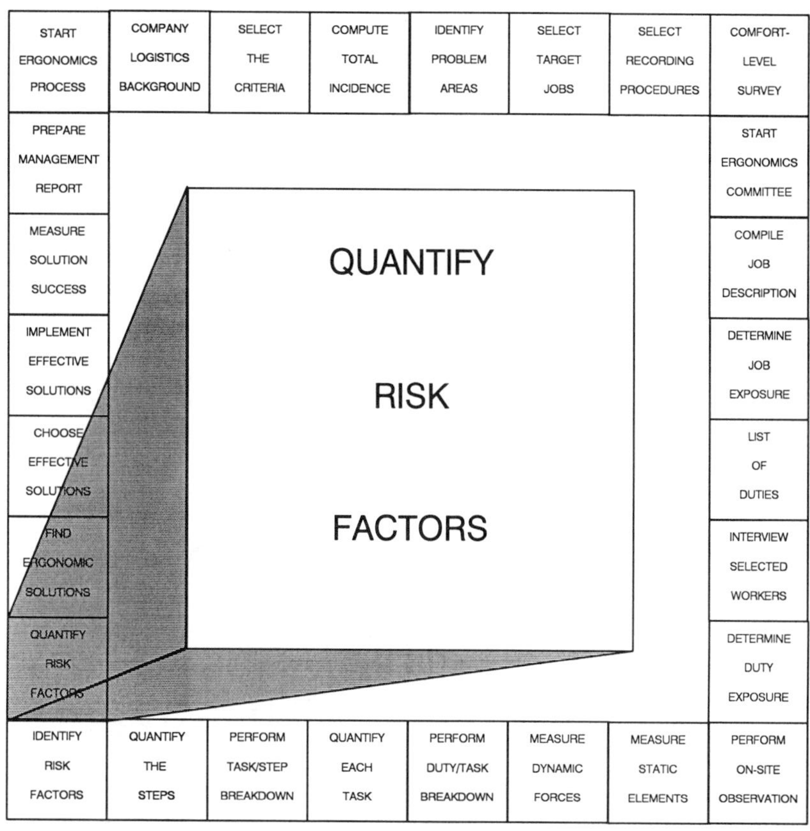
START ERGONOMICS PROCESS
COMPANY LOGISTICS BACKGROUND
SELECT THE CRITERIA
COMPUTE TOTAL INCIDENCE
IDENTIFY PROBLEM AREAS
SELECT TARGET JOBS
SELECT RECORDING PROCEDURES
COMFORT-LEVEL SURVEY
START ERGONOMICS COMMITTEE
COMPILE JOB DESCRIPTION
DETERMINE JOB EXPOSURE
LIST OF DUTIES
INTERVIEW SELECTED WORKERS
DETERMINE DUTY EXPOSURE
PERFORM ON-SITE OBSERVATION
MEASURE STATIC ELEMENTS
MEASURE DYNAMIC FORCES
PERFORM DUTY/TASK BREAKDOWN
QUANTIFY EACH TASK
PERFORM TASK/STEP BREAKDOWN
QUANTIFY THE STEPS
IDENTIFY RISK FACTORS
QUANTIFY RISK FACTORS
FIND ERGONOMIC SOLUTIONS
CHOOSE EFFECTIVE SOLUTIONS
IMPLEMENT EFFECTIVE SOLUTIONS
MEASURE SOLUTION SUCCESS
PREPARE MANAGEMENT REPORT
QUANTIFY
RISK
FACTORS

Quantify Risk Factors

OBJECTIVE

To list and examine the identified risk factors.

GUIDELINES

There are several ways to use the initial results of your risk identification survey. You might simply list all the risk factors and identify them according to job titles. This can be helpful in quantifying the more challenging or hazardous jobs. This can then be compared to the current injury records. This information can be a helpful component in an early return-to-work or light-duty program. You might also use this breakdown as a method for prioritizing the efforts to try to solve the problems in the area.

Some other possible ways to examine the data include the following. You might list all the risk factors and next to each one enter the number of times each one was identified. Also, you might list each risk factor and identify with which job titles, duties, or tasks it was associated.

There can be some value in listing risk factors according to a body part. This information can also be correlated with the company history of injuries of cumulative trauma disorder and used to prioritize the needs for progressing with the intervention discovery process. This step can be considered to be elective.

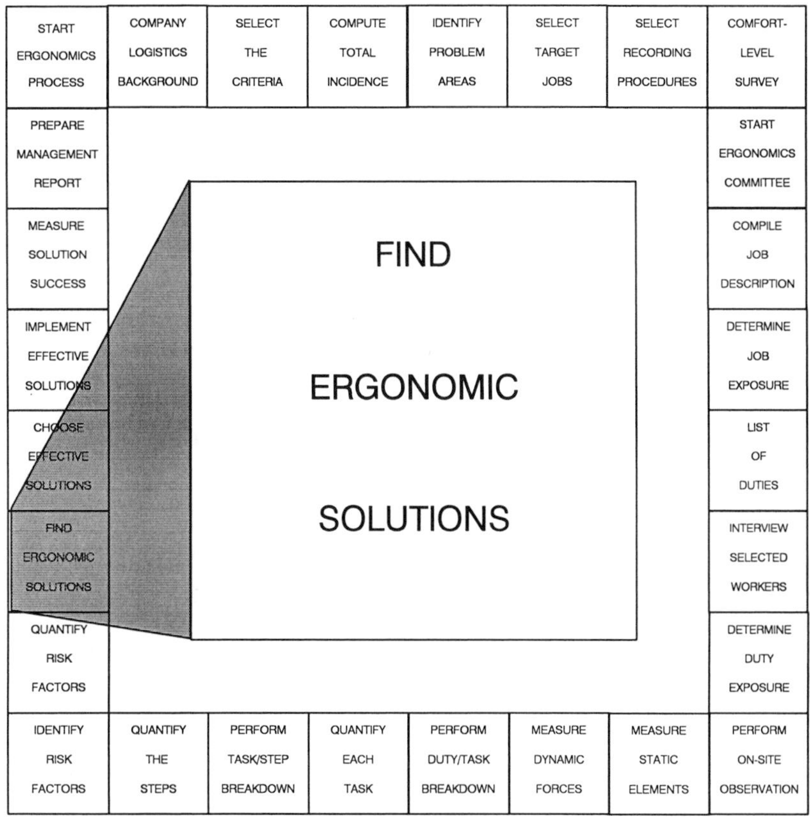
START ERGONOMICS PROCESS
COMPANY LOGISTICS BACKGROUND
SELECT THE CRITERIA
COMPUTE TOTAL INCIDENCE
IDENTIFY PROBLEM AREAS
SELECT TARGET JOBS
SELECT RECORDING PROCEDURES
COMFORT-LEVEL SURVEY
START ERGONOMICS COMMITTEE
COMPILE JOB DESCRIPTION
DETERMINE JOB EXPOSURE
LIST OF DUTIES
INTERVIEW SELECTED WORKERS
DETERMINE DUTY EXPOSURE
PERFORM ON-SITE OBSERVATION
MEASURE STATIC ELEMENTS
MEASURE DYNAMIC FORCES
PERFORM DUTY/TASK BREAKDOWN
QUANTIFY EACH TASK
PERFORM TASK/STEP BREAKDOWN
QUANTIFY THE STEPS
IDENTIFY RISK FACTORS
QUANTIFY RISK FACTORS
FIND ERGONOMIC SOLUTIONS
CHOOSE EFFECTIVE SOLUTIONS
IMPLEMENT EFFECTIVE SOLUTIONS
MEASURE SOLUTION SUCCESS
PREPARE MANAGEMENT REPORT
FIND
ERGONOMIC
SOLUTIONS

Find Ergonomic Solutions

OBJECTIVE

To compose a comprehensive list of possible interventions to improve working conditions, to enhance performance, and to facilitate consistent quality by decreasing the effects of each of the risk factors identified.

SUPPLIES

1. List of risk factors
2. Intervention discovery guide

The goal of this step is to arrive at a comprehensive list of possible interventions to decrease the effects of each of the risk factors identified. You will need to consider the various strategies for reducing each risk factor.

Ergonomics is strongly based on common sense. Many of the truly effective interventions are ones that people will react to by saying, "Why didn't I think of that?" It is certainly not an exact science. While there are many situations where pure logic can dictate an answer, there are instances where the opposite of what is intuitive is more effective. One example would be if you were trying to hear a person speaking to you at a noisy gathering. Believe it or not, you will hear them better if you stick your fingers in your ears.

When considering interventions, creativity and unabashed brainstorming become essential skills to be utilized. The most preposterous idea may well have a hint of possibility in it, or it may stimulate an idea that will significantly reduce the problem.

There is no "one" absolute, correct answer. There are numerous effective solutions. The challenge is to consider as many solutions as possible in an organized and comprehensive manner to find the ones most appropriate to your organization. Because there are so many possible solutions, this task can be overwhelming. The following model is an attempt to organize the numerous ergonomic intervention options into smaller, more practical categories.

Most interventions can be placed into one of four categories. They are:

Input
Output
Machine
Worker

The *input* is the item or items to be processed, moved, or acted upon in any way by a specific worker. The *output* is the final condition of that input as a result of the actions taken by the worker during the task being studied. The *machine* is the tangible concrete object(s) to which the worker will be exposed. This would include the workstation, tools, carts, environmental conditions, and many other items. The *worker* is just that, the "human" worker.

Example

In a cookie factory, the cookie icer is responsible for icing cookies. The input are the yet-to-be-iced cookies as they pass by on the conveyor belt. The output are the iced cookies sitting on a tray and placed in the rack behind the worker. The bowl, spatula, conveyor, and chair are all part of the machine, and of course, the cookie icer is the worker.

For a cross-country truck driver, the input would be the truck located in New York. His output would be the truck located in Los Angeles. The cab of the truck is the machine.

INPUT/OUTPUT CONSIDERATIONS

Start by looking at a particular task or duty and determine the input and output. Once you have identified the input, consider how making a change would alter the risk factors associated with the activities being performed. Typical considerations here are the position of the input as presented; the rate at which it is presented; and the size, shape, weight, and even temperature of the input.

Ask yourself or discuss in a group such questions as "In order to decrease this particular risk factor..."

How could you change the size of the container/item?

How could you order the materials in a different-sized box?

How could you have the supplier perform part of the task?

If packaged in-house, how could you have it delivered in a different-sized box?

Try not to judge any of the interventions prematurely. The process of screening them will come in the next step.

Consider the output. Determine what the output is in the particular duty, task, or step being examined. Then consider what effect would occur if changes were made.

"In order to reduce the risk factors in this step..."

Could you make output rate variable?

Could you not act upon the input in any way by shipping with "some assembly required?"

Focusing on just these two aspects of the process is extremely important. This will help you avoid the tendency to focus only on mechanical interventions, thus limiting simpler and less expensive options.

WORKER INTERVENTION CONSIDERATIONS

Next, consider ways that a worker-based program can contribute to decreasing the effects of an unnecessary stress. This is where education programs, setting procedural policies, exercise programs, and worker selection are considered. These can be a little more difficult to visualize. Additional information to expand upon the goals and components of various worker interventions can be found in Appendix D.

The intervention discovery guide at the end of this chapter will help in the consideration of the many options in this area.

MACHINE INTERVENTION CONSIDERATIONS

The last major category to be considered deals with the machinery and environment. As this area has the greatest number of variables, it is important to systematically assess the various modifications. It is important to look at the problem from several different perspectives. First, consider what act is being performed. Is something being moved? Is something being held in place?

For example, consider what can be done to adjust the work. This would include ideas such as adjusting work heights on workstations, installing a lift table, splitting workstation heights, tilting bins towards the worker, or installing a turntable device.

In some cases it may not be practical to adjust the work or workstation. The work may be too large or too delicate to be adjusted, or someone may have just purchased brand-new nonadjustable workstations. Therefore, you should consider what can be done to adjust the worker. This would include providing adjustable chairs, or sit/stand chairs, placing the worker on a platform or gantry to raise him, or perhaps in a subfloor, such as the pit a mechanic works in to lower himself.

If something is being moved, you should consider such options as specialized carts, overhead gantries, forklifts, and conveyor belts. If, on the other hand, something is being held in place, the appropriate options are installation of a tool balancer, design of a jig, or use of an overhead arm for positioning.

Increasing accessibility deals primarily with bringing things to the worker with less stress. This can be something as simple as cutting down the front of a cardboard box that holds parts or tilting a conveyor belt towards a worker. A big part of this strategy involves the use of specialized tools or extensions on existing tools.

Supporting worker body parts in a way that is the least stressful is perhaps one of the more overlooked and simple considerations. This deals with the design of seating, foot rests, and adding supports such as low-back cushions on existing chairs or armrests that extend from the workstation. Another consideration has to do with augmenting sensory input. This can involve modification to controls, repositioning display terminals or speakers, adding adjustable lighting, or providing special glasses.

Along with the idea of sensory input, you should consider the controls. Such possible interventions include changing buttons to levers, levers to foot pedals, placing an extension on a lever, simply making the button to be pushed larger, or moving it to a more convenient location.

The next consideration involves thinking about ways of protecting the worker's body from hard, sharp, forceful, heat, cold, or vibration exposure. This can be as simple as adding padding, using temporary handles, or providing gloves.

Since many of the risk factors deal with performing a stressful motion repetitively, it can be helpful to focus specifically on this issue. Interventions that are likely to arise from this type of approach would be storing tools or materials in a better location, use of a specialized tool, automation of a process, or, when possible, partially automating the process. This can be something as simple as providing an electric can opener instead of a manual one.

Thus far many of the approaches have dealt with tools or equipment. It is critical that those tools and pieces of equipment be maintained. Many risk factors can be reduced by examining the method currently used to maintain working order or sharpness of tools. Setting specific criteria for needs and providing a simple reporting mechanism can help in dealing with these considerations.

A worker's perception of the environment can affect his overall performance. Whenever possible, a worker should have some control over the environmental conditions in his area. This may be providing him with personal heaters or fans, or sometimes just providing him with a thermometer to assure him that his subjective assessment is not a true indicator of the conditions. In the cases where a thermometer is provided, it is important to have a simplified reporting mechanism to convey information regarding heating, ventilating, and air-conditioning considerations.

The intervention discovery guide is a series of algorithms that help to prompt you through various types of interventions. It does not contain every possible intervention. It is designed to help stimulate other ideas that may be more appropriate for your situation.

INTERVENTION DISCOVERY PROCEDURES

Start with your list of risk factors. Determine if you are going to look at all the risk factors at once; look at them in groups according to duty, task, or step; or if you can, consider each risk factor separately. This decision will be based on practical considerations, such as time, experience, and the total number of risk factors identified in your analysis.

1. List the risk factors and where they occurred by duty, task, or step.
2. Look at page 1 of the intervention discovery guide. Follow the instructions regarding going to additional pages to consider each of the options suggested, or for more specific information.
3. Write down all possible interventions on a worksheet similar to the intervention discovery form.
4. When the various questions have been exhausted for a particular risk factor or group of risk factors, move onto the next group and start at page 1 and repeat the process.

You can use the intervention forms to generate your initial worksheet list. The report format introduced in the risk identification step can now be expanded (see Appendix G).

INTERVENTIONS WORKSHEET

Page *1* of *1*

Duty *Filling orders*

Task *Pulling stock*

1. Place frequently used items on lower shelves.

2. Provide step stools for getting items from high shelves.

3. Provide reacher devices to get small items from high shelves.

4. Initiate warm-up exercise program for low back and shoulders.

5. Provide shock-absorbing floor mats.

6. Provide shock-absorbing insoles.

7. Provide training to workers covering the safest possible material-handling postures and practices.

8. Provide lift tables for handling heavy, awkward loads.

9. Provide automatic tape-dispensing equipment.

INTERVENTION DISCOVERY GUIDE

Intervention Discovery 

Guide 1

The input is the item or items to be processed, moved, or acted upon in any way by a specific worker.

1. Can you change or alter the input in any way?

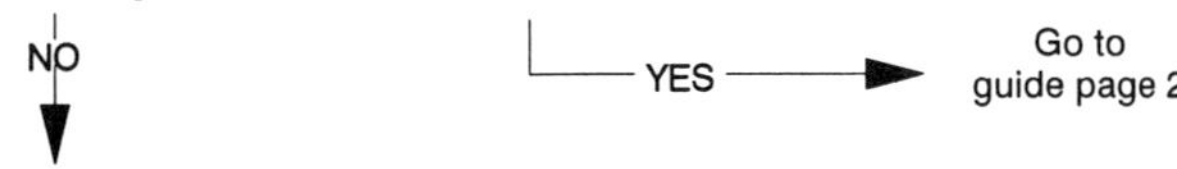

Go to Next Question

The output is the final product. The output is the final condition of the input as a result of the action taken.

2. Can you change or alter the output in any way?

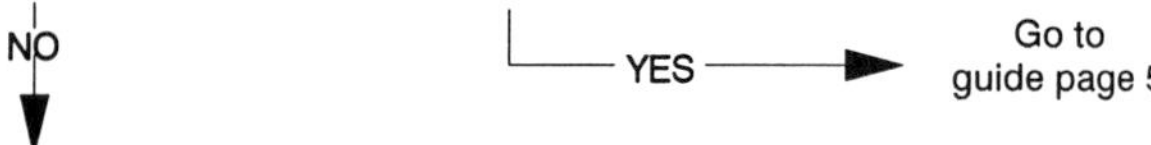

Go to guide page (6)

(Page numbers are within this guide only.)

ALG-PRO1

Intervention Discovery 

Guide 2

Would any of the following decrease this particular risk factor?

1. Change the size of the container/item.

2. Order the materials in a different sized box.

3. Have the supplier perform part of the task.

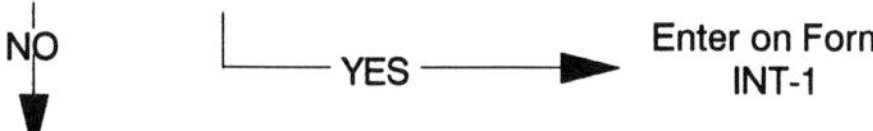

4. If packaged in-house, have it delivered in a different format.

5. Increase QA to maintain more rigid standards for input materials.

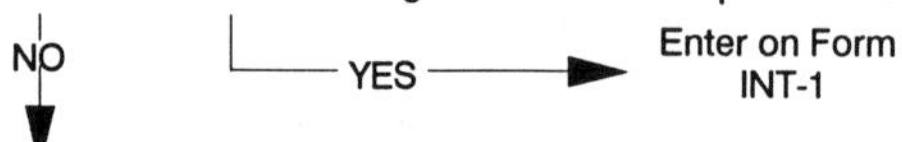

6. Have materials delivered to a different area.

7. Increase the weight of the container/item.

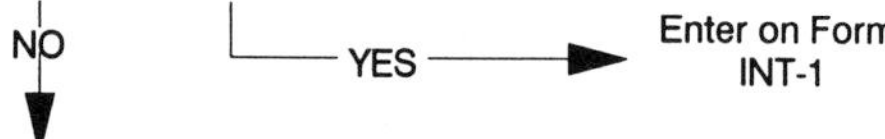

8. Change the shape of the input.

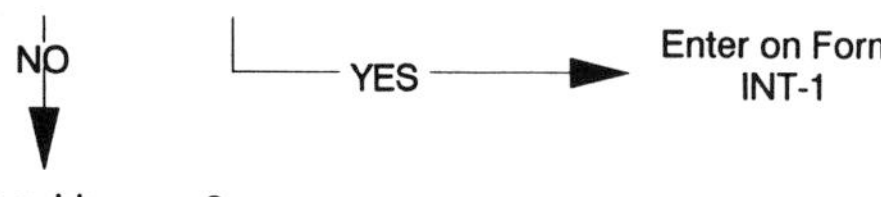

ALG-IN1a

Intervention Discovery 

Guide 3

Would any of the following decrease this particular risk factor?

1. Decrease the weight of the container.

2. Change the number of different pieces to be handled.

3. Change the composition of the materials.

4. Alter the properties of the materials (malleability, heat conductivity, etc.).

5. Alter environmental conditions that affect material properties.

6. Change the way or position in which the input is presented.

7. Change the container in which the materials come.

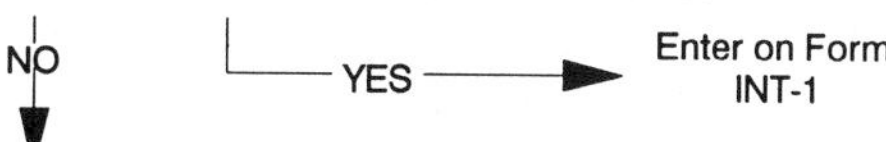

8. Break the tasks into component parts.

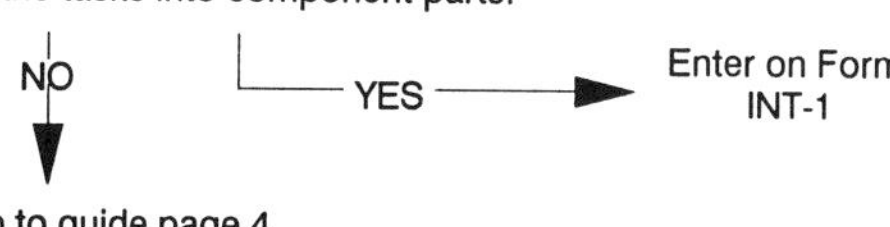

Go to guide page 4

ALG-IN1b

Intervention Discovery

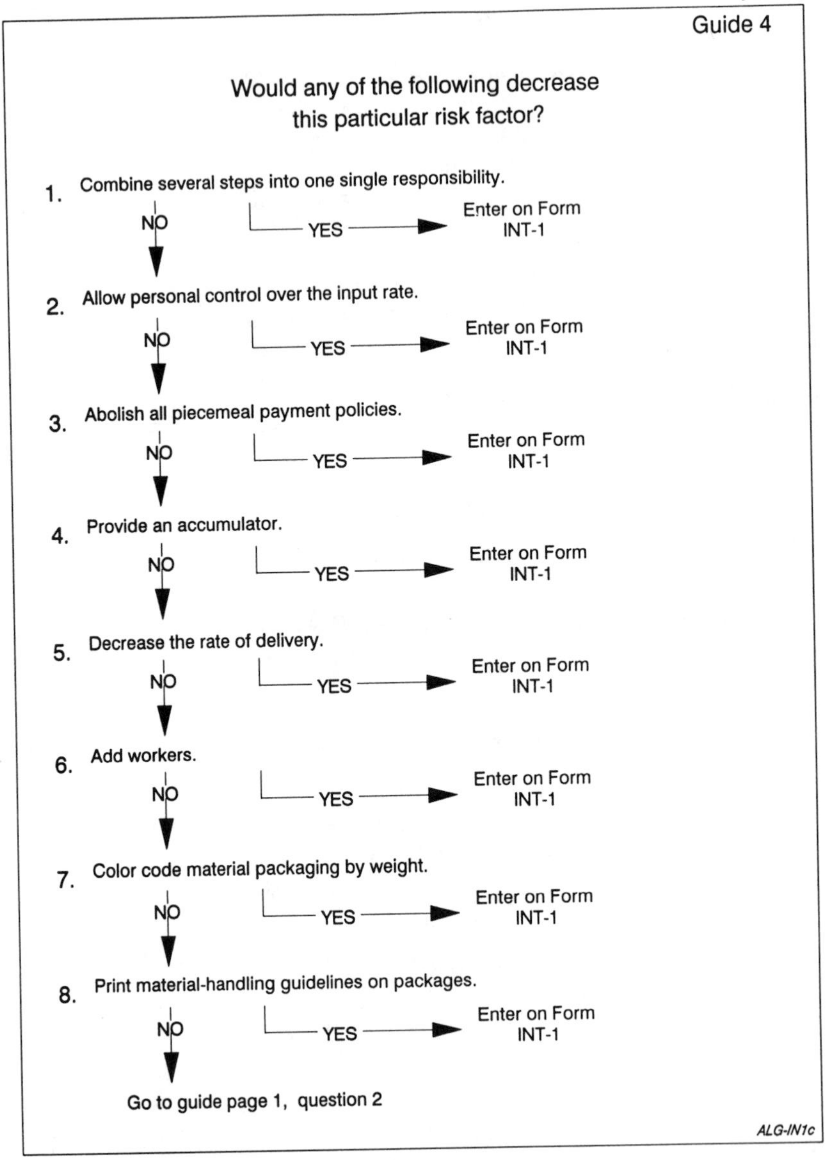

Guide 4

Would any of the following decrease this particular risk factor?

1. Combine several steps into one single responsibility.
   NO ↓ — YES → Enter on Form INT-1
2. Allow personal control over the input rate.
   NO ↓ — YES → Enter on Form INT-1
3. Abolish all piecemeal payment policies.
   NO ↓ — YES → Enter on Form INT-1
4. Provide an accumulator.
   NO ↓ — YES → Enter on Form INT-1
5. Decrease the rate of delivery.
   NO ↓ — YES → Enter on Form INT-1
6. Add workers.
   NO ↓ — YES → Enter on Form INT-1
7. Color code material packaging by weight.
   NO ↓ — YES → Enter on Form INT-1
8. Print material-handling guidelines on packages.
   NO ↓ — YES → Enter on Form INT-1

Go to guide page 1, question 2

ALG-IN1c

Intervention Discovery 

Guide 5

Would any of the following decrease this particular risk factor?

1. Increase or decrease the number of steps performed by each worker.

2. Have the worker pack or put away whatever he or she processes.

3. Hire more workers

4. Make the output rate variable.

5. Don't act upon the input (ship as is).

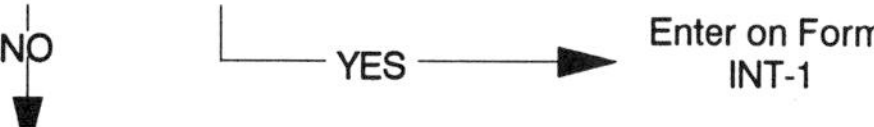

6. Assure that the level of quality is necessary for the customer.

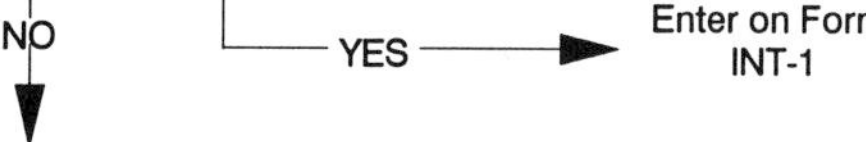

7. Stop using the currently used storage space.

8. Deliver the product with "Some assembly required".

Go to guide page 6, question 1

ALG-OU1a

Intervention Discovery 

WORKER INTERVENTIONS

Guide 6

1. Can this risk factor be reduced by initiating an education or training program? — YES → Go to guide page 7

NO ↓

2. Can this risk factor be reduced by mandating procedures? — YES → Go to guide page 9

NO ↓

3. Can you initiate an exercise program? — YES → Go to guide page 10

NO ↓

4. Can you utilize personal protective equipment? — YES → Go to guide page 11

NO ↓

5. Can you utilize monitoring devices? — YES → Go to guide page 12

NO ↓

6. Can you initiate rotation or light duty programs? — YES → Go to guide page 13

NO ↓

7. Can you utilize a worker selection process? — YES → Go to guide page 14

NO → Go to guide page 15

ALG-WORK

Intervention Discovery 

Guide 7

What specific information should be included in this training?

1. Activities to do at home to decrease the risk
2. Exercises to be performed at home
3. Exercises to be performed at work
4. Home activities that can increase the risk
5. How to adjust the equipment
6. How to adjust the seating
7. How to adjust the work station
8. How to communicate the need for maintenance
9. How to maintain tools
10. How to perform stretching exercises all day
11. How to recognize the need for maintenance
12. Least stressful work practices (lifting, job-specific procedures)
13. Mandated procedures
14. Need for proper-fitting clothes
15. Need for proper-fitting personal protective equipment
16. Principles of ergonomics
17. Performance methods
18. Postures to avoid

List the content of the proposed training along with the suggested format from guide page 8 on Form INT-1.

Go to guide page 8

ALG-WK1a

Intervention Discovery

Guide 8

What training format would be effective?

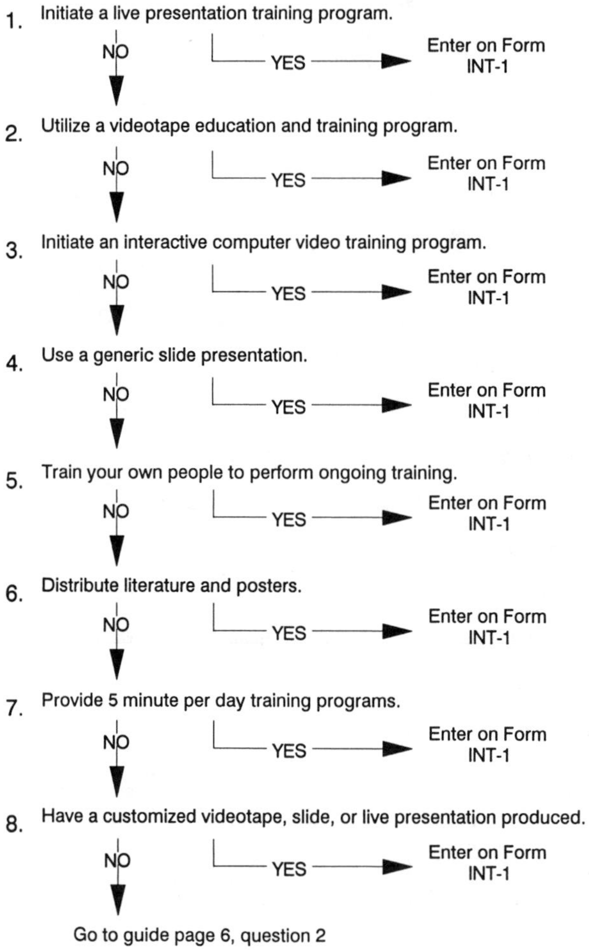

1. Initiate a live presentation training program.
   NO ↓ — YES → Enter on Form INT-1
2. Utilize a videotape education and training program.
   NO ↓ — YES → Enter on Form INT-1
3. Initiate an interactive computer video training program.
   NO ↓ — YES → Enter on Form INT-1
4. Use a generic slide presentation.
   NO ↓ — YES → Enter on Form INT-1
5. Train your own people to perform ongoing training.
   NO ↓ — YES → Enter on Form INT-1
6. Distribute literature and posters.
   NO ↓ — YES → Enter on Form INT-1
7. Provide 5 minute per day training programs.
   NO ↓ — YES → Enter on Form INT-1
8. Have a customized videotape, slide, or live presentation produced.
   NO ↓ — YES → Enter on Form INT-1

Go to guide page 6, question 2

ALG-WK1b

Intervention Discovery 

Guide 9

Would any of the following decrease this particular risk factor?

1. Hold each worker accountable for group performance.

2. Hold on-line management accountable for group performance.

3. Set up a demerit system leading to suspension or lay off.

4. Reward compliance with incentive programs.

Return to guide page 6, question 3

ALG-WK2

Intervention Discovery 

Guide 10

Would any of the following decrease this particular risk factor?

1. Initiate upper extremity flexibility and strengthening warm up exercises.

2. Initiate low back flexibility and strengthening warm up exercises.

3. Initiate an overall flexibility and strengthening exercise program.

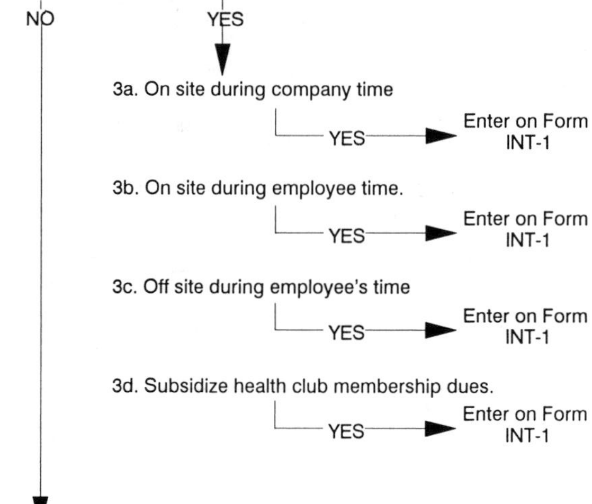

Return to guide page 6, question 4

ALG-WK3

Intervention Discovery 

Guide 11

Would any of the following decrease this particular risk factor?

1. Have workers wear wrist splints.

2. Have workers wear back braces.

3. Have workers wear lift belts.

Return to guide page 6, question 5

ALG-WK4

Intervention Discovery 

Guide 12

Would any of the following decrease this particular risk factor?

1. Utilize a monitoring device to assess the number of repetitions of the action.

2. Use a device to provide feedback when performing a task in a stressful way.

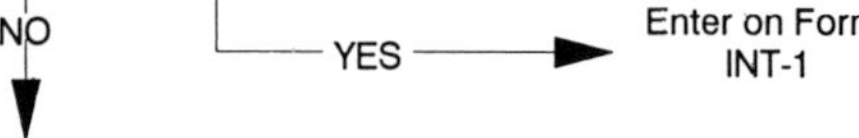

Return to guide page 6, question 6

ALG-WK5

Intervention Discovery 

Guide 13

Would any of the following decrease this particular risk factor?

1. Initiate a light-duty/heavy duty rotation program.

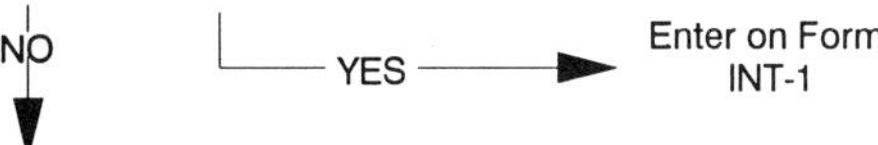

2. Initiate work / rest cycles.

3. Initiate weekly rotation to jobs using different muscle groups.

4. Initiate daily rotation to jobs using different muscle groups.

5. Initiate daily periodic rotation to jobs using different muscle groups.

6. Initiate VALID light-duty program for injured workers.

7. Initiate light-duty / break-in period for new workers.

8. Initiate light-duty / break-in period for transferees.

Return to guide page 6, question 7

ALG-WK6

Intervention Discovery 

Guide 14

Would any of the following decrease this particular risk factor?

1. Utilize strength testing in worker selection or placement process.

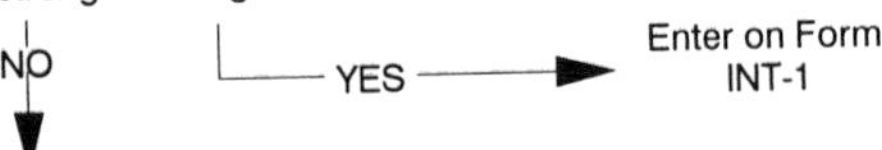

2. Utilize flexibility testing in worker selection or placement process.

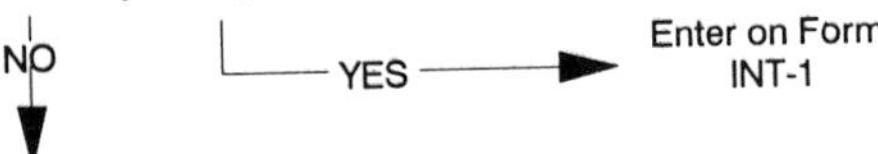

3. Utilize medical record review in worker selection or placement process.

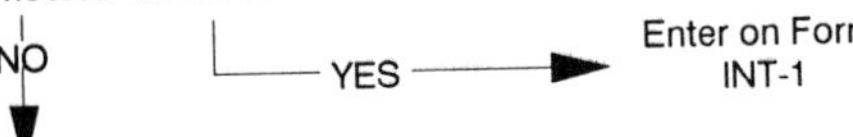

4. Utilize functional capacity testing in worker selection or placement process.

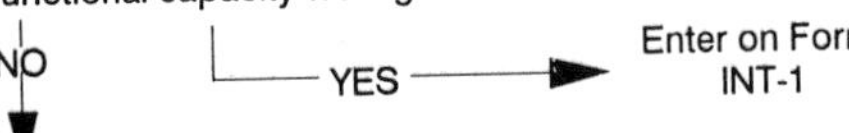

5. Utilize physical fitness testing in worker selection or placement process.

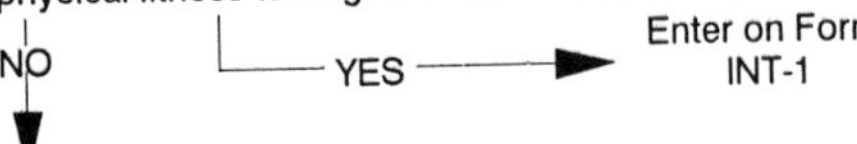

6. Utilize vibrometry testing in worker selection or placement process.

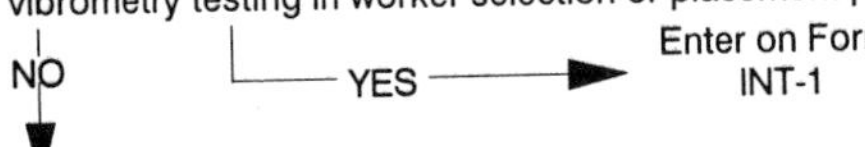

7. Utilize neurometry testing in worker selection or placement process.

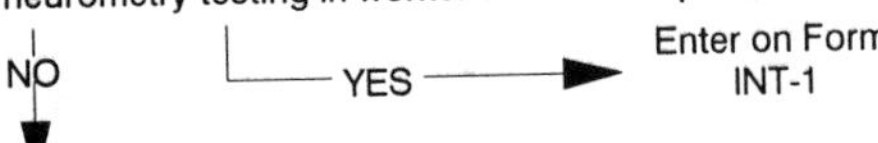

8. Utilize sensation testing in worker selection or placement process.

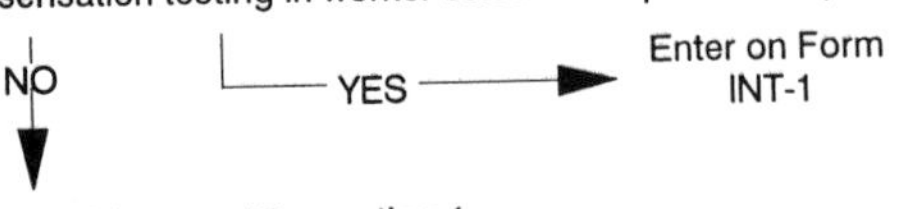

Go to guide page 15, question 1

ALG-WK7

Intervention Discovery 

MACHINE INTERVENTIONS

Guide 15

1. Can this risk factor be reduced by adjusting the position of the worker? — YES → Go to guide page 17

NO ↓

2. Can this risk factor be reduced by adjusting the position of the workstation? — YES → Go to guide page 18

NO ↓

3. Is something being moved? — YES → Go to guide page 21

NO ↓

4. Is something being held in place? — YES → Go to guide page 25

NO ↓

5. Does this involve repetitive activity? — YES → Go to guide page 27

NO ↓

6. Is the worker unsupported in any way? — YES → Go to guide page 28

NO ↓

7. Does this involve restricted access? — YES → Go to guide page 30

NO → Go to next page

ALG-MAC1

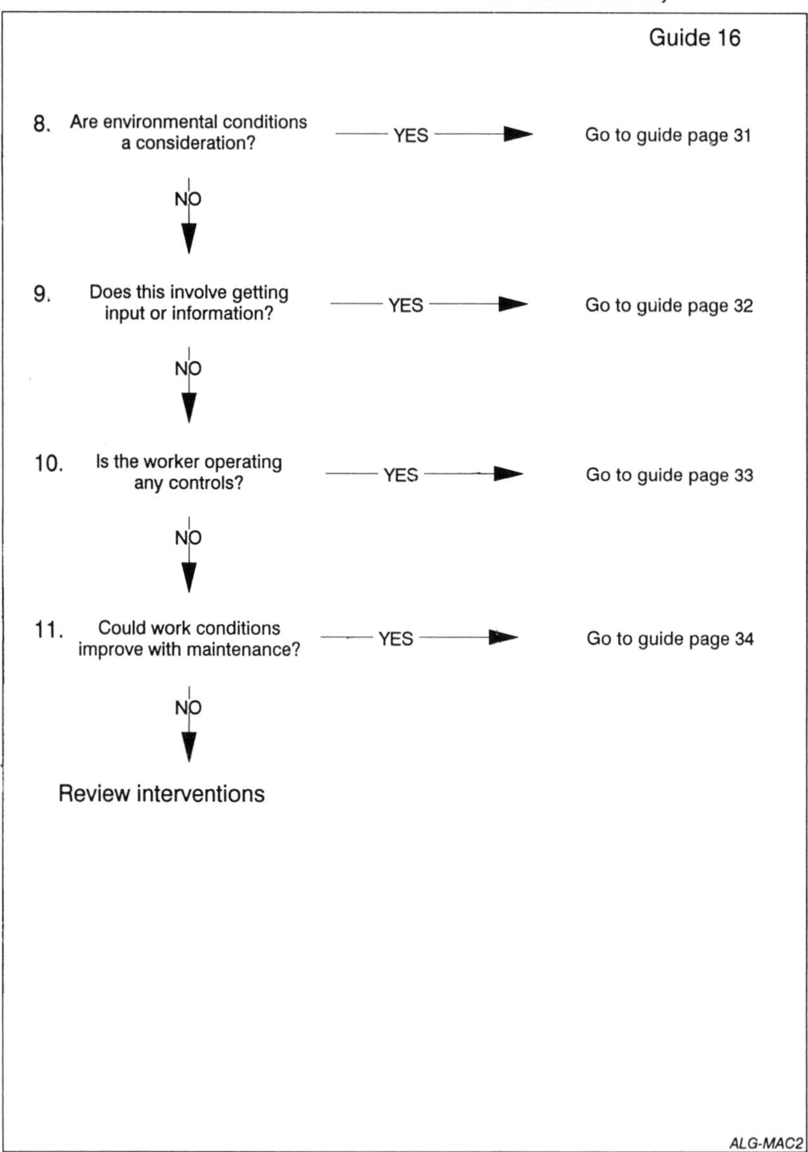
Intervention Discovery
Guide 16
8. Are environmental conditions a consideration?
YES
Go to guide page 31
NO
9. Does this involve getting input or information?
YES
Go to guide page 32
NO
10. Is the worker operating any controls?
YES
Go to guide page 33
NO
11. Could work conditions improve with maintenance?
YES
Go to guide page 34
NO
Review interventions
ALG-MAC2

Intervention Discovery 

Guide 17

Would any of the following decrease this particular risk factor?

1. Adjust chair height, angle, depth, or armrest position.

2. Replace current chairs with adjustable models.

3. Place the worker on a platform to raise him/her.

4. Place worker in a trough or sunken floor to lower him/her.

5. Place worker on pallets

6. Provide adjustable floor height.

7. Provide sit/stand chair.

8. Use fixed scaffolding or gantry.

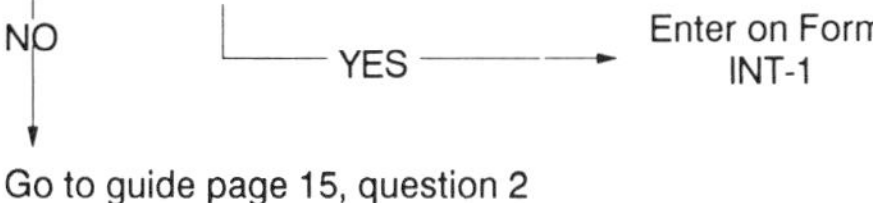

ALG-MA1

Intervention Discovery 

Guide 18

Would any of the following decrease this particular risk factor?

1. Adjust cart, table, or workstation height.

2. Replace current carts, tables, or workstations with adjustable models.

3. Place the workstation or collection bin on a platform.

4. Place work surface or collection bin in a trough or sunken floor to lower.

5. Place work station on Pallets.

6. Attach work surface or collection bin to the wall at the appropriate height.

7. Place sleeves or extenders on levers.

8. Place extenders on workstation or table legs.

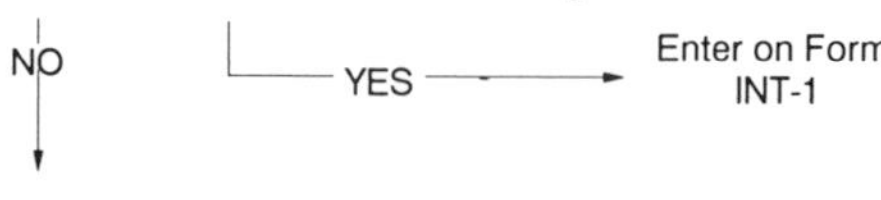

Go to guide page 19

ALG-MA2a

Intervention Discovery

Guide 19

Would any of the following decrease this particular risk factor?

1. Split work surface into two or more heights.

2. Use a spring-loaded, double bottom or self-adjusting cart or bin.

3. Place a short platform on the top of the work surface.

4. Place a cushion on the chair.

5. Angle work surface towards the worker by raising the back legs.

6. Angle towards the worker by placing an easel or tilted platform on work surface.

7. Angle towards the worker by mounting on the wall.

8. Change the angle of the conveyor.

Go to guide page 20

ALG-MA2b

Intervention Discovery 

Guide 20

Would any of the following decrease this particular risk factor?

1. Install lazy Susan or turntable to turn or tilt work.

2. Cut out work surface to allow the worker to get closer.

3. Change multiple work surface layout from 180 to 90°.

4. Place chair on track.

5. Place item on swinging arm.

6. Add third hand or document holder.

Go to guide page 15, question 3

ALG-MA2c

Intervention Discovery 

Guide 21

Would any of the following decrease this particular risk factor?

1. Is a hand tool being used in this process?
   - NO ↓
   - YES → Go to guide page 35

2. Can a hand tool be used in any part of this process?
   - NO ↓
   - YES → Go to guide page 37

3. Make changeable pieces easier to change (slip on vs screw on).
   - NO ↓
   - YES → Enter on Form INT-1

4. Install lazy susan to rotate and/or tilt work as appropriate.
   - NO ↓
   - YES → Enter on Form INT-1

5. Decrease rolling resistance of carts, carriers, or conveyance devices.
   - NO ↓
   - YES → Enter on Form INT-1

6. Utilize customized or redesigned hand trucks.
   - NO ↓
   - YES → Enter on Form INT-1

7. Utilize a special applications dolly (barrel, plywood, etc.).
   - NO ↓
   - YES → Enter on Form INT-1

8. Install a cart-loading mechanism.
   - NO ↓
   - YES → Enter on Form INT-1

Go to guide page 22

ALG-MA3a

Intervention Discovery 

Guide 22

Would any of the following decrease this particular risk factor?

1. Install special attachments for jacks.

2. Make trays or bins out of lighter material.

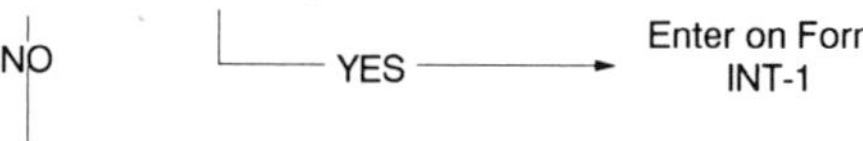

3. Utilize an overhead gantry, boom, or fixed hoist.

4. Place collection ports or areas lower to utilize gravity.

5. Place ramps over thresholds.

6. Install temporary handles.

7. Install permanent handles.

8. Install a feeder mechanism.

Go to guide page 23

ALG-MA3b

Intervention Discovery 

Guide 23

Would any of the following decrease
this particular risk factor?

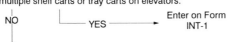

1. Place multiple shelf carts or tray carts on elevators.

2. Place rollers at the edge of the table.

3. Use a conveyor belt.

4. Use a forklift.

5. Use a portable lifting device.

6. Use a rolling lever or portable rollers (for cylinders, wire/paper spools).

7. Use a suction lift device.

8. Use a two-man stretcher device (two pole suspension).

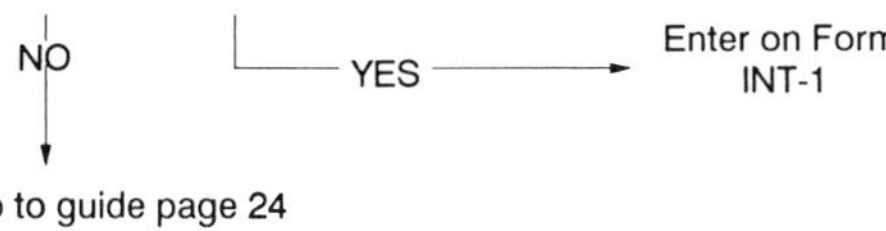

ALG-MA3c

Intervention Discovery 

Guide 24

Would any of the following decrease this particular risk factor?

1. Use a vacuum tube system to move objects.

2. Use an air gun or nozzle device.

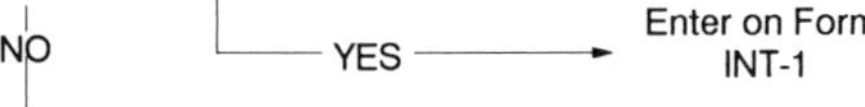

3. Use an overhead (tube fed) dispenser mechanism.

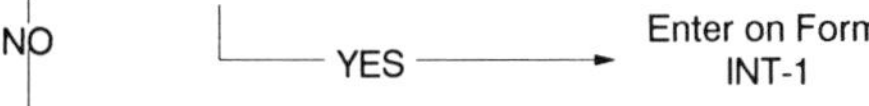

4. Use slip sheeting.

NO — YES → Enter on Form INT-1

5. Use an oversized forklift.

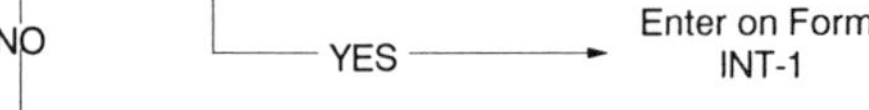

6. Use a lever type (fishing pole) device.

NO — YES → Enter on Form INT-1

7. Use lubricants to decrease the resistance to movement.

8. Change the wheels on the carts.

Go to guide page 15, question 4

ALG-MA3d

Intervention Discovery 

Guide 25

Would any of the following decrease this particular risk factor?

1. Is a hand tool being used in this process?
   NO ↓ — YES → Enter on Form INT-1

2. Can a hand tool be used in any part of this process?
   NO ↓ — YES → Enter on Form INT-1

3. Use a customized or stock jig.
   NO ↓ — YES → Enter on Form INT-1

4. Install special attachment for jacks.
   NO ↓ — YES → Enter on Form INT-1

5. Install a vise for holding in place.
   NO ↓ — YES → Enter on Form INT-1

6. Add a weight to lower part of the item.
   NO ↓ — YES → Enter on Form INT-1

7. Use a forklift.
   NO ↓ — YES → Enter on Form INT-1

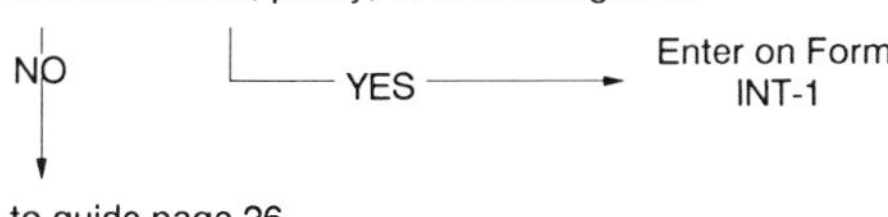

8. Use an overhead hoist, pulley, or articulating arm.
   NO ↓ — YES → Enter on Form INT-1

Go to guide page 26

ALG-MA4a

Intervention Discovery 

Guide 26

Would any of the following decrease this particular risk factor?

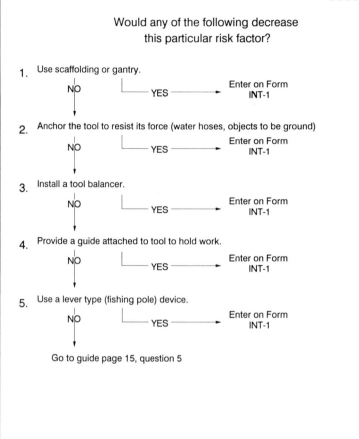

1. Use scaffolding or gantry.

NO — YES → Enter on Form INT-1

2. Anchor the tool to resist its force (water hoses, objects to be ground)

NO — YES → Enter on Form INT-1

3. Install a tool balancer.

NO — YES → Enter on Form INT-1

4. Provide a guide attached to tool to hold work.

NO — YES → Enter on Form INT-1

5. Use a lever type (fishing pole) device.

NO — YES → Enter on Form INT-1

Go to guide page 15, question 5

ALG-MA4b

Intervention Discovery 

Guide 27

Would any of the following decrease this particular risk factor?

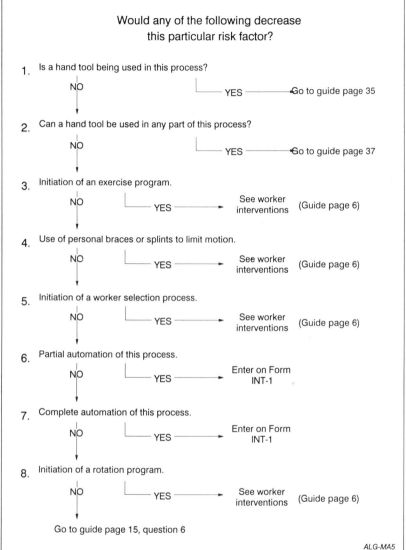

1. Is a hand tool being used in this process?
   - NO ↓
   - YES → Go to guide page 35

2. Can a hand tool be used in any part of this process?
   - NO ↓
   - YES → Go to guide page 37

3. Initiation of an exercise program.
   - NO ↓
   - YES → See worker interventions (Guide page 6)

4. Use of personal braces or splints to limit motion.
   - NO ↓
   - YES → See worker interventions (Guide page 6)

5. Initiation of a worker selection process.
   - NO ↓
   - YES → See worker interventions (Guide page 6)

6. Partial automation of this process.
   - NO ↓
   - YES → Enter on Form INT-1

7. Complete automation of this process.
   - NO ↓
   - YES → Enter on Form INT-1

8. Initiation of a rotation program.
   - NO ↓
   - YES → See worker interventions (Guide page 6)

Go to guide page 15, question 6

ALG-MA5

Intervention Discovery

Guide 28

Would any of the following decrease this particular risk factor?

1. Provide a chair.
   NO ↓ — YES → Enter on Form INT-1

2. Install a bar or step stool under the workstation.
   NO ↓ — YES → Enter on Form INT-1

3. Place a chair on ladder or gantry for extended work.
   NO ↓ — YES → Enter on Form INT-1

4. Provide an adjustable footrest.
   NO ↓ — YES → Enter on Form INT-1

5. Provide adjustable armrests.
   NO ↓ — YES → Enter on Form INT-1

6. Provide chairs that allow height, tilt, and backrest variability.
   NO ↓ — YES → Enter on Form INT-1

7. Place a recess into the tables to support arms.
   NO ↓ — YES → Enter on Form INT-1

8. Bevel all workstation edges and railings.
   NO ↓ — YES → Enter on Form INT-1

Go to guide page 29

ALG-MA6a

Intervention Discovery 

Guide 29

Would any of the following decrease this particular risk factor?

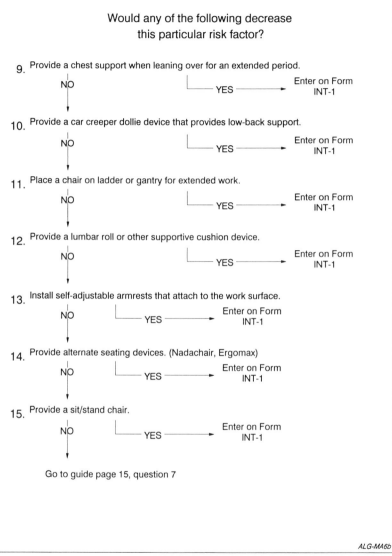

9. Provide a chest support when leaning over for an extended period.

NO → YES → Enter on Form INT-1

10. Provide a car creeper dollie device that provides low-back support.

NO → YES → Enter on Form INT-1

11. Place a chair on ladder or gantry for extended work.

NO → YES → Enter on Form INT-1

12. Provide a lumbar roll or other supportive cushion device.

NO → YES → Enter on Form INT-1

13. Install self-adjustable armrests that attach to the work surface.

NO → YES → Enter on Form INT-1

14. Provide alternate seating devices. (Nadachair, Ergomax)

NO → YES → Enter on Form INT-1

15. Provide a sit/stand chair.

NO → YES → Enter on Form INT-1

Go to guide page 15, question 7

ALG-MA6b

Intervention Discovery 

Guide 30

Would any of the following decrease this particular risk factor?

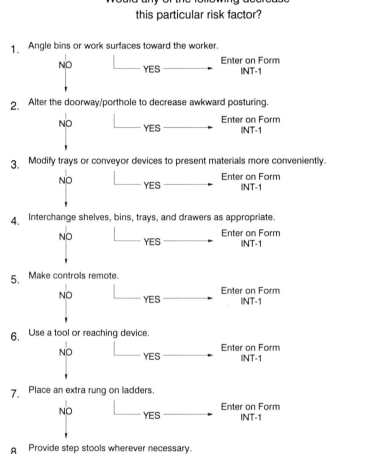

1. Angle bins or work surfaces toward the worker.

2. Alter the doorway/porthole to decrease awkward posturing.

3. Modify trays or conveyor devices to present materials more conveniently.

NO — YES → Enter on Form INT-1

4. Interchange shelves, bins, trays, and drawers as appropriate.

5. Make controls remote.

6. Use a tool or reaching device.

7. Place an extra rung on ladders.

8. Provide step stools wherever necessary.

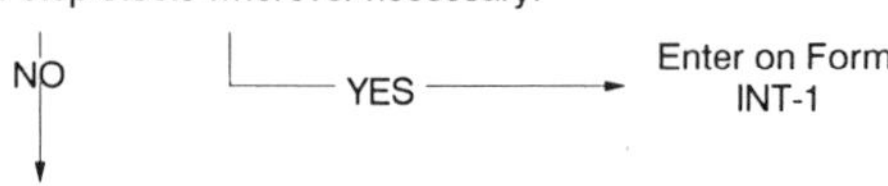

Return to guide page 16, question 8

ALG-MA7

Intervention Discovery 

Guide 31

Would any of the following decrease this particular risk factor?

1. Give worker personal control over heating or air conditioning.

2. Provide a reporting mechanism for H.V.A.C. questions/concerns.

3. Provide personal heating or cooling devices at workstations (heaters, fans).

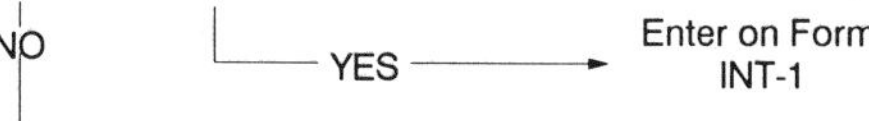

4. Check all climatic monitoring systems for accuracy.

5. Install insulating dividers in areas of greatest climatic variability.

6. Call in H.V.A.C. consultant.

7. Provide comfortable protective clothing.

8. Allow greater variability in worker's attire (sweaters, tee shirts).

Return to guide page 16, question 9

ALG-MA8

Intervention Discovery 

Guide 32

Would any of the following decrease this particular risk factor?

1. Move meters, visual displays, or other input device to more accessible space.

2. Install adjustable lighting.

3. Provide headphones for audio input.

4. Provide lighted magnifying glass.

5. Provide glare screen or reposition workstation to decrease glare.

6. Increase size of dials or visual display equipment.

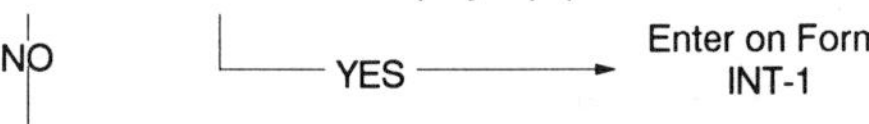

7. Provide personal adjustability in location and display features.

8. Provide ear plugs.

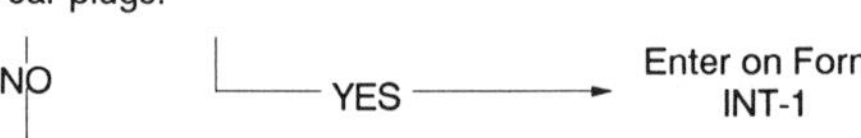

Return to guide page 16, question 10

ALG-MA9

Intervention Discovery 

Guide 33

Would any of the following decrease this particular risk factor?

1. Make handles for different functions in different shapes.
   NO ↓ — YES → Enter on Form INT-1

2. Make control handles/buttons larger.
   NO ↓ — YES → Enter on Form INT-1

3. Place sleeves or extenders on difficult to reach levers/controls.
   NO ↓ — YES → Enter on Form INT-1

4. Provide personal choice in hand or foot control selection.
   NO ↓ — YES → Enter on Form INT-1

5. Make foot or hand controls moveable for personal placement.
   NO ↓ — YES → Enter on Form INT-1

6. Interchange buttons, levers, dials, and smooth or beveled knobs as appropriate.
   NO ↓ — YES → Enter on Form INT-1

7. Move all overhead controls to a more accessible position.
   NO ↓ — YES → Enter on Form INT-1

8. Place controls inset into the table for more upper extremity support.
   NO ↓ — YES → Enter on Form INT-1

Return to guide page 16, question 11

ALG-MA10

Intervention Discovery 

Guide 34

Would any of the following decrease this particular risk factor?

1. Establish a process for requesting maintenance of tools or work station.
   NO ↓ — YES → Enter on Form INT-1

2. Set up simple rules for identifying the need for maintenance.
   NO ↓ — YES → Enter on Form INT-1

3. Establish a regular maintenance schedule.
   NO ↓ — YES → Enter on Form INT-1

4. Initiate equipment inspection program to assess the need for maintenance.
   NO ↓ — YES → Enter on Form INT-1

5. Establish a flooring (rugs, surface coverings) inspection program.
   NO ↓ — YES → Enter on Form INT-1

6. Initiate inspection program for tool sharpness.
   NO ↓ — YES → Enter on Form INT-1

7. Initiate inspection program for tool accuracy/calibration.
   NO ↓ — YES → Enter on Form INT-1

8. Initiate inspection program to make sure safety features are intact.
   NO ↓ — YES → Enter on Form INT-1

Review Form INT-1

ALG-MA11

Intervention Discovery 

Guide 35

Would any of the following decrease this particular risk factor?

1. Change tool handle design by altering the angle.

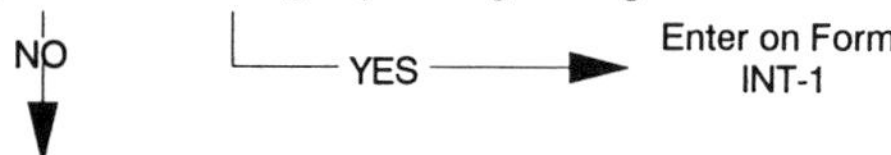

2. Change handle design by adding padding.

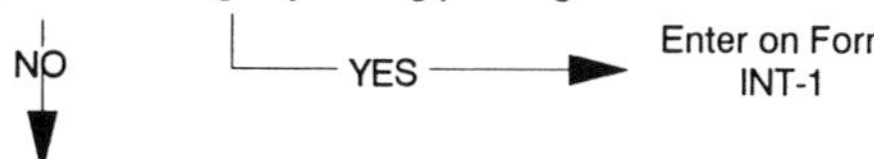

3. Change handle design by altering the tool shape.

4. Change tool design by altering the size, diameter, weight, or length.

5. Simplify tool adjustment.

6. Use several fixed tools rather than one adjustable one.

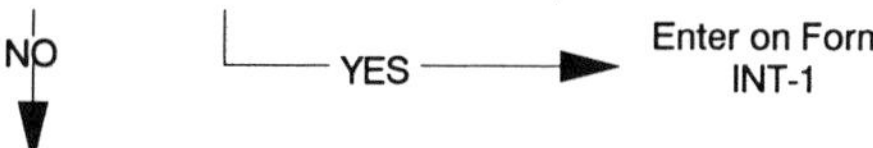

7. Make tools electric, air, or battery operated.

8. Store tools in more convenient position.

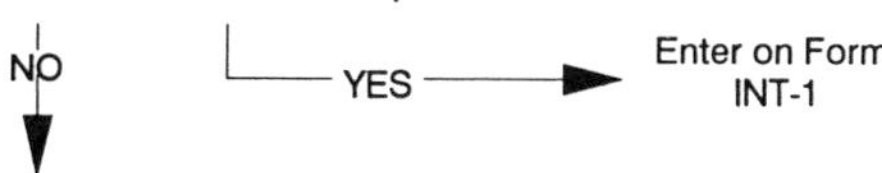

Go to guide page 36

ALGMAT1a

Intervention Discovery

Guide 36

Would any of the following decrease this particular risk factor?

1. Install a tool balancer.
   NO ↓ / YES → Enter on Form INT-1

2. Utilize robotics.
   NO ↓ / YES → Enter on Form INT-1

3. Automate part of the process.
   NO ↓ / YES → Enter on Form INT-1

4. Automate the entire process.
   NO ↓ / YES → Enter on Form INT-1

Go to guide page 21, question 2

ALGMAT1b

Intervention Discovery

Guide 37

Would any of the following decrease this particular risk factor?

1. Utilize a hammer.
   NO ↓ | YES → Enter on Form INT-1

2. Utilize a wrench.
   NO ↓ | YES → Enter on Form INT-1

3. Utilize pliers.
   NO ↓ | YES → Enter on Form INT-1

4. Utilize a screwdriver or nut driver.
   NO ↓ | YES → Enter on Form INT-1

5. Utilize a reacher.
   NO ↓ | YES → Enter on Form INT-1

6. Utilizer a pop riveter or stapler.
   NO ↓ | YES → Enter on Form INT-1

7. Utilize a glue gun.
   NO ↓ | YES → Enter on Form INT-1

8. Utilize a hose and nozzle.
   NO ↓ | YES → Enter on Form INT-1

Go to guide page 38

ALGMAT2a

Intervention Discovery 

Guide 38

Would any of the following decrease this particular risk factor?

1. Use straps or bars to handle bulky, awkward loads.

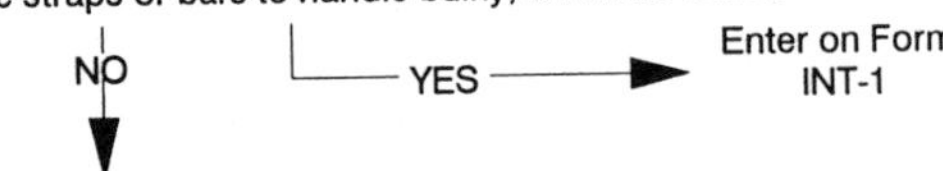

2. Use temporary handles.

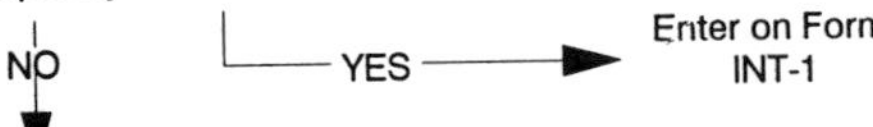

3. Convert a fixed tool on the workstation into a hand tool.

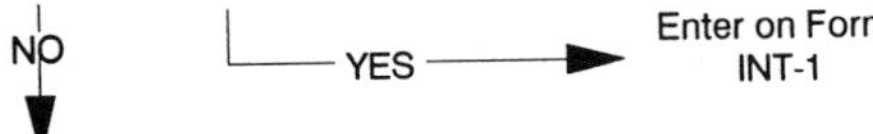

4. Make a hand tool into a fixed tool.

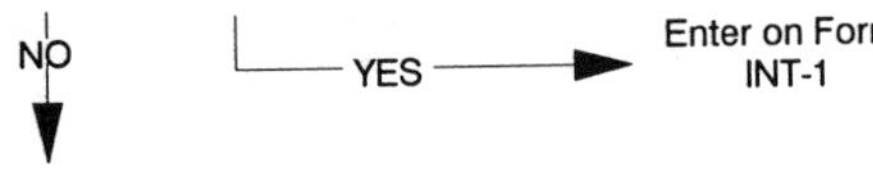

Go to guide page 21, question 3.

ALGMAT2b

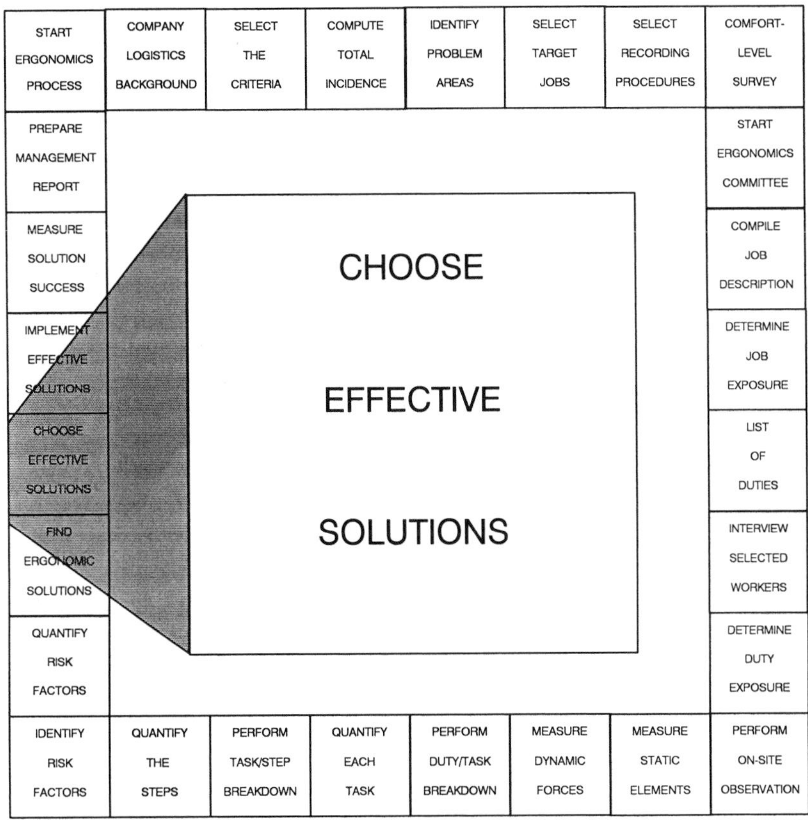

START ERGONOMICS PROCESS
COMPANY LOGISTICS BACKGROUND
SELECT THE CRITERIA
COMPUTE TOTAL INCIDENCE
IDENTIFY PROBLEM AREAS
SELECT TARGET JOBS
SELECT RECORDING PROCEDURES
COMFORT-LEVEL SURVEY
START ERGONOMICS COMMITTEE
COMPILE JOB DESCRIPTION
DETERMINE JOB EXPOSURE
LIST OF DUTIES
INTERVIEW SELECTED WORKERS
DETERMINE DUTY EXPOSURE
PERFORM ON-SITE OBSERVATION
MEASURE STATIC ELEMENTS
MEASURE DYNAMIC FORCES
PERFORM DUTY/TASK BREAKDOWN
QUANTIFY EACH TASK
PERFORM TASK/STEP BREAKDOWN
QUANTIFY THE STEPS
IDENTIFY RISK FACTORS
QUANTIFY RISK FACTORS
FIND ERGONOMIC SOLUTIONS
CHOOSE EFFECTIVE SOLUTIONS
IMPLEMENT EFFECTIVE SOLUTIONS
MEASURE SOLUTION SUCCESS
PREPARE MANAGEMENT REPORT
CHOOSE EFFECTIVE SOLUTIONS

Choose Effective Solutions

OBJECTIVE

To determine the most effective and practical interventions from the preliminary list of possibilities.

SUPPLIES

1. List of possible interventions

GUIDELINES

In this step you will take a look at the various interventions and determine which ones to implement. There are any number of criteria to be used for this screening process. The following will describe just a few of them and discuss their potential effectiveness. The goal of the screening mechanism is to choose the most effective intervention and the most effective area in which to implement it. In most cases the very first step is to determine the cost.

1. What is the cost in dollars of the equipment or materials necessary for implementation?
2. What is the cost of the personnel necessary to install the equipment?
3. Will this implementation in any way delay the production? If so, what is the cost of that delay?
4. What is the cost of time the worker will spend learning to utilize the modification?

SCREENING BY AREA

If you are considering many different areas in your facility, refer back to the criteria you used to determine which area demonstrated the greatest need for ergonomic intervention. Next, examine all the possible interventions considered for that area and predict the outcome of their implementation.

Next, look at the second greatest area of need, look at all the interventions available there, and predict the outcome. The natural question is, "Why bother looking at anything other than the area of greatest need?" The answer is that you are trying to be maximally effective. This will assist in attaining future support for ergonomic programs and interventions. It may also become apparent that the interventions in the second greatest area of need will cost less and yet save more money.

Let's say the estimated cost of injuries in the packaging area is $100,000, and the only practical interventions are projected to decrease injuries by 20%. In the inspection area, the cost of injuries is $50,000, and you estimate that your intervention will result in a 50% reduction in injury incidence. There is a greater anticipated return by implementing an intervention in the secondary area.

SUBJECTIVE CRITERIA

In an ideal world you could look at the success of past companies and determine which particular intervention has been proven to be effective. While that type of background study is strongly recommended, it is still going to come down to a rather subjective decision.

No two plants are exactly the same. Even if the process and layout were exactly the same, the people who make up the workforce in each of those facilities are going to differ. Because of this, you cannot be totally objective in your prediction of the outcome of a particular intervention.

In some cases, it may just be a matter of picking the one that seems the easiest to implement, get accepted, and be complied with.

ORIGINATING GUIDELINES

In spite of this lack of objective data relative to the effectiveness of particular interventions, certain rationales can assist in the decisionmaking process.

Rationale 1: Productivity and Performance vs. Comfort and Safety

This model is an attempt to put a numerical assessment on the projected effectiveness based on the anticipated impact upon worker productivity and performance, and worker comfort and safety.

An intervention is examined in fairly practical terms and rated according to the following point score. If several managers are involved then the points can be averaged.

One statement from each group is chosen.

A. This intervention if implemented would

5 Definitely increase worker comfort and safety
4 Possibly increase worker comfort and safety
3 Result in no change in worker comfort and safety
2 Possibly decrease worker comfort and safety
1 Definitely decrease worker comfort and safety

B. This intervention if implemented would

5 Definitely increase productivity and performance
4 Possibly increase productivity and performance
3 Result in no change in productivity and performance
2 Possibly decrease productivity and performance
1 Definitely decrease productivity and performance

The points from the top column are multiplied by the points from the bottom column. The best intervention is one that will definitely increase productivity and performance and also will definitely increase worker comfort and safety.

This would result in a maximum score of 25 when multiplied together. A form is included at the end of this section to assist in this process.

Rationale 2: Projected Impact

Another method is to assess the relative projected effectiveness of different types of interventions. To make this simpler, assign a number to that projected effectiveness. For instance, if you decide to hire an independent contractor to perform a step of an operation that has been identified as being a significant risk for developing knee problems, then you have actually eliminated that risk. In other words, subcontracting a problem is 100% effective. If, on the other hand, you decide to take a wait-and-see attitude and choose to do nothing, the projected effectiveness would be 0%.

These numerical estimations may be utilized independently or in combination with rationale 1 outlined above. The effectiveness percent you assign to each intervention is going to be subjective at first. Over time, with careful tracking, you should be able to arrive at your own specific effectiveness projections.

Examples of projected effectiveness criteria are

Total Cost of Injuries $1,000,000

	Estimated Effectiveness	
Intervention	Percent	Cost
Elimination	10%	$100,000
Education programs	5	50,000
Worker selection	15	150,000
Engineering	25 (after costs)	250,000

The estimates table above represents the prediction that the elimination of a particular risk factor will result in a 10% reduction in injuries. It states that the initiation of an education program will result in an overall reduction in injuries of 5%. The numbers cited are provided as examples and are not meant to provide universal guidelines.

Rationale 3: Cost

Cost is often a consideration and can be added to the formula. You may want to break the various ergonomic options into groups of $500 or less, $501–2000, and greater than $2000. You can look at this information independently or in combination with the other rationales. The ideal intervention is the one projected to have the greatest positive impact on productivity and worker safety, to have the greatest effectiveness, and to cost the least.

Rationale 4: Management Commitment

Another extremely important variable is the amount of management commitment that is necessary. In situations where the management team is firmly behind an ergonomics effort, this is less of a concern. In some cases, you may be dealing with a group that still needs to be shown how effective this can be. In that case interventions requiring less management commitment will have a better chance of succeeding.

As you can see, there is still a great deal of subjectivity. It is important to note that no one program is going to solve all your ergonomic injury problems. It is going to take a comprehensive approach. Only through careful tracking of rationale for intervention selection and the resulting effectiveness can you develop your own database for future decisions.

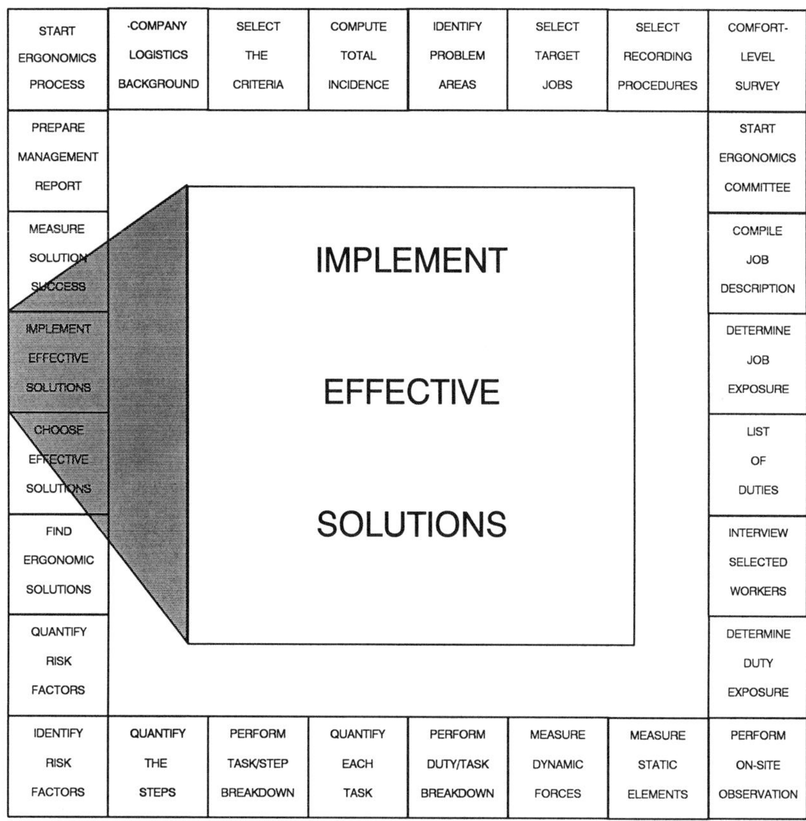
START ERGONOMICS PROCESS
·COMPANY LOGISTICS BACKGROUND
SELECT THE CRITERIA
COMPUTE TOTAL INCIDENCE
IDENTIFY PROBLEM AREAS
SELECT TARGET JOBS
SELECT RECORDING PROCEDURES
COMFORT-LEVEL SURVEY
START ERGONOMICS COMMITTEE
COMPILE JOB DESCRIPTION
DETERMINE JOB EXPOSURE
LIST OF DUTIES
INTERVIEW SELECTED WORKERS
DETERMINE DUTY EXPOSURE
PERFORM ON-SITE OBSERVATION
MEASURE STATIC ELEMENTS
MEASURE DYNAMIC FORCES
PERFORM DUTY/TASK BREAKDOWN
QUANTIFY EACH TASK
PERFORM TASK/STEP BREAKDOWN
QUANTIFY THE STEPS
IDENTIFY RISK FACTORS
QUANTIFY RISK FACTORS
FIND ERGONOMIC SOLUTIONS
CHOOSE EFFECTIVE SOLUTIONS
IMPLEMENT EFFECTIVE SOLUTIONS
MEASURE SOLUTION SUCCESS
PREPARE MANAGEMENT REPORT
IMPLEMENT
EFFECTIVE
SOLUTIONS

Implement Effective Solutions

OBJECTIVE

To develop an organized and systematic approach to implementing the chosen interventions.

GUIDELINES

The steps to initiating an intervention are equally as important as determining which intervention to implement. It is important to utilize an organized and systematic approach to find the product or service that will meet the specific need of the risk factor you are hoping to affect. There are certain steps to be followed:

1. Plan how to implement.
2. Train and orient employees.
3. Install the new intervention.
4. Collect initial feedback.

PROCEDURES

Planning

The procedures will depend on the type of intervention to be implemented.

If the intervention is a product, a hard good, or a piece of hardware, then it is relatively easy.

1. Contact various vendors.
2. Request information.
3. Review information.
4. Meet with suppliers.
5. Ask questions.

The following list is not meant to be comprehensive, nor to replace any procedures currently in place for the procurement of supplies. It should act as a reminder that purchasing a product or service for ergonomic intervention should be viewed as critically as you would if you were purchasing a device aimed at increasing productivity.

1. What are the names of customers who have placed this device in a similar setup as yours?

2. For companies that have had success with this product, were there any other interventions initiated at the same time?
3. For companies that have had success with this device, were there any other significant changes in work force, production, or processes during this time?
4. Has there ever been an increase in injuries as a result of this product?
5. Is there an additional charge for late-night or weekend installation?
6. Is there an additional charge for training?
7. What is the final cost?
8. How long does it take to get delivery?
9. How difficult is this to install?
10. How long will it take to actually install?
11. How has the installation impacted productivity in the past?
12. Can the installation be done during normal down time, such as late-night or weekends?
13. How much training is necessary?
14. Do they do the training themselves or is it subcontracted?
15. What about the availability of support immediately following the implementation?

Contact others who have used this device for information. Try to confirm the effectiveness claims provided by the vendors. Be wary of all research. Much of the research is anecdotal. It is typically the results of the recordkeeping efforts of either the vendor or the person who was responsible for the use of the intervention at a facility. Either one of those sources would benefit from a successful outcome of the research. Look to research performed by totally independent sources. In most cases these will come from academic institutions where strict adherence to the scientific method is employed. The most critical component is the utilization of a control group to assess the effectiveness of an intervention.

If the intervention is an education program or a worker selection process, the selection can be even more challenging than for a product, a hard good, or a piece of hardware. It may be helpful to look through Appendix D, which addresses the various worker-based programs.

In the case of an education program, you may want to attend a presentation. This may mean that you sit in on another company's program. If this is not possible, you can ask to see a videotape of an "actual" presentation. The quality of the videotape will not be important. You should be more interested in the interaction of the presenter with the audience, the content of the materials, and the industry specificity. After viewing this presentation, try to answer the following questions:

Is it specific to your industry?

Is it interesting?

Can it be performed in an entertaining way?

Can retention be tested?

Is there a follow-up program to reinforce the program?

Does it take into account activities outside of the workplace?

Does it discuss the concept of personal responsibility?

Does it provide some definable and useful tools?

Does it address the specific risk factor that precipitated this intervention?

Will it give the participants something to talk about?

Can it be performed in a reasonable amount of time?

What is the expected rate of compliance?

Are there any support materials left behind?

Worker Selection

The goal of a worker selection process is to fit the physical attributes of a job candidate to a task for which he or she is capable or to determine the existence of conditions that may cause physical harm to the candidate in performing normal job duties. There is nothing wrong with asking to see copies of the procedures to be used. In addition, you should see a copy of the report.

Is it job specific?

Does it follow the guidelines of the EEOC, the Rehabilitation Act of 1973, the current Civil Rights Act, and the Americans with Disabilities Act?

Is there anything that needs to be done before initiating a worker selection process?

What is being tested?

Is it safe?

Has anyone ever been injured during the performance of this test?

Has anyone ever challenged the validity of the test in a court? What was the resolution of that case?

Who will be responsible for determining the criteria for pass/fail?

Is this a functional capacity test or a diagnostic procedure?

Does the information in the report tell you what you want to know?

How long will it take to schedule a test (48-hour minimum)?

How long will it take to get a verbal report?

How long will it take to get a written report?

This is a rather controversial strategy to take to reduce injuries. Be certain to have explored all the legal ramifications of taking this approach prior to initiating this process.

The goal of a worker conditioning or exercise program is to help the worker better meet the physical demands of the job. This is achieved by making it more convenient for the worker to perform strengthening, flexibility, or cardiovascular exercises. These programs may be aimed at strengthening particular muscle groups to aid in performance of a job, warm-up exercises, or overall cardiovascular conditioning.

1. What are the names of customers who have utilized this program in a similar setup as yours?
2. For companies that have had success with this program, were there any other interventions initiated at the same time?
3. For companies that have had success with this program, were there any other significant changes in work force, production, or processes during this time?
4. Has there ever been an increase in injuries as a result of this product?
5. Does this require on-going supervision?
6. Can it be done independently?
7. Are there additional materials and costs?
8. If there is equipment involved, what is the anticipated life of that equipment?
9. Can worker performance of these programs be monitored?

Training

No one likes to be told what to do. Change can also meet with resistance. One of the big challenges of initiating an ergonomics intervention is to "sell" the idea to the individuals who will be affected by it. This will certainly be the individuals who will be using it, but it is equally important to communicate with anyone dependent on this area for a set production rate or quality assurance standards. The new intervention may suddenly interrupt old standards. There may suddenly be an increase or decrease in the rate of production. Individuals who are going to be affected need to know why this is happening.

Training in new methods should answer the following questions:

Why this intervention?

Why now?

Why in this area?

When will this start?

What are the expected outcomes?

How this will affect them?

What is the mechanism for feedback?

Can this be changed back?

How long will it be tried?

Who to ask for help or additional training?

Once you have covered all that information, you can cover this last question.

How should the new intervention be utilized?

It is preferable that each individual be given the opportunity to try the new device or intervention. This is best facilitated by a nonstressful, unchallenged environment.

Installation or Initiation

Installation should be performed as per agreement by a qualified technician working with an hourly worker, if at all possible. This will give the worker the opportunity to learn, as well as being on hand to recognize any conditions that may not have been taken into consideration before (e.g., did you know that it blocks the fire exit?)

Initial testing by the installer is a must. Have the installer actually operate all the functions of the intervention. Have someone from your facility also operate all the functions of the intervention before the installing technician leaves.

Collect Initial Feedback

There will always be questions during the first few days of a new process. Either an in-house person should be designated to field those questions or a representative from the provider should be on hand. This will significantly reduce the stress that is likely to be generated and therefore ensure maximum compliance.

It may be helpful to also collect some additional feedback a week or so after the initial installation. At the very least, this will confirm that the new intervention is being utilized. This is also necessary if an education program was provided. It is helpful to find out if techniques discussed in the presentation are being utilized.

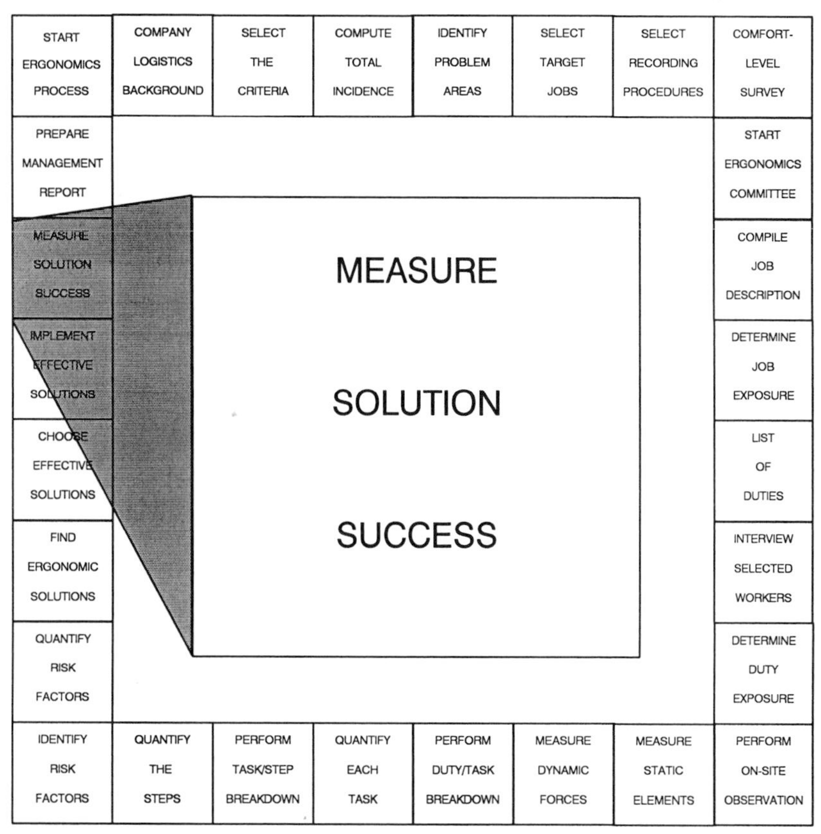
START ERGONOMICS PROCESS
COMPANY LOGISTICS BACKGROUND
SELECT THE CRITERIA
COMPUTE TOTAL INCIDENCE
IDENTIFY PROBLEM AREAS
SELECT TARGET JOBS
SELECT RECORDING PROCEDURES
COMFORT-LEVEL SURVEY
START ERGONOMICS COMMITTEE
COMPILE JOB DESCRIPTION
DETERMINE JOB EXPOSURE
LIST OF DUTIES
INTERVIEW SELECTED WORKERS
DETERMINE DUTY EXPOSURE
PERFORM ON-SITE OBSERVATION
MEASURE STATIC ELEMENTS
MEASURE DYNAMIC FORCES
PERFORM DUTY/TASK BREAKDOWN
QUANTIFY EACH TASK
PERFORM TASK/STEP BREAKDOWN
QUANTIFY THE STEPS
IDENTIFY RISK FACTORS
QUANTIFY RISK FACTORS
FIND ERGONOMIC SOLUTIONS
CHOOSE EFFECTIVE SOLUTIONS
IMPLEMENT EFFECTIVE SOLUTIONS
MEASURE SOLUTION SUCCESS
PREPARE MANAGEMENT REPORT
MEASURE SOLUTION SUCCESS

Measure Solution Success

OBJECTIVE

To assess the effectiveness of the ergonomics intervention efforts.

GUIDELINES

The first step is to look at the initial criteria selected for measurement of the effectiveness. Compare the objective data that have been collected. Next, qualify these results. This is done by considering as many factors as possible that may have positively or negatively affected the results. It may seem foolish to test your results if they came out positive. However, if the intervention utilized is to be expanded to other areas, you want to be as certain as possible that the results are reproducible.

There is a science to statistical analysis that, while it might be ideally suited for this type of exercise, is not practical for the scope of this book. A more feasible process would be to answer the following questions.

Did the desired and anticipated effect of the intervention occur?

Did any other benefits result from this intervention?

Did any negative effects occur?

Was there compliance with the intervention?

If there was a partial compliance with the intervention, was there a proportional desired effect? (Example: Half the workers utilized the new lift device and the result was a 20% reduction in injuries instead of the anticipated 40%.)

Was the tracking mechanism followed?

Was any additional information necessary?

Was this the only intervention implemented during this time?

Were the reporting schedules followed?

Was all the follow-up carried out?

Was there management support?

Were there any changes in the management, the equipment, the work force, or the process that may have affected the outcome of this intervention?

What was the monetary gain or loss as a result of this intervention?

Did it work?

PROCEDURES

Once you are certain of the level of validity of your data, the next few steps follow as a natural progression.

1. Compute the changes in the baseline data to the measure the outcome of the intervention.
2. Complete an additional evaluation form to validate the results of the data.
3. Compile a report for upper management.
4. If a control group was used, perform a comparative analysis.
5. State the level of success or failure.
6. Provide an action plan to either correct or expand on this or other interventions.

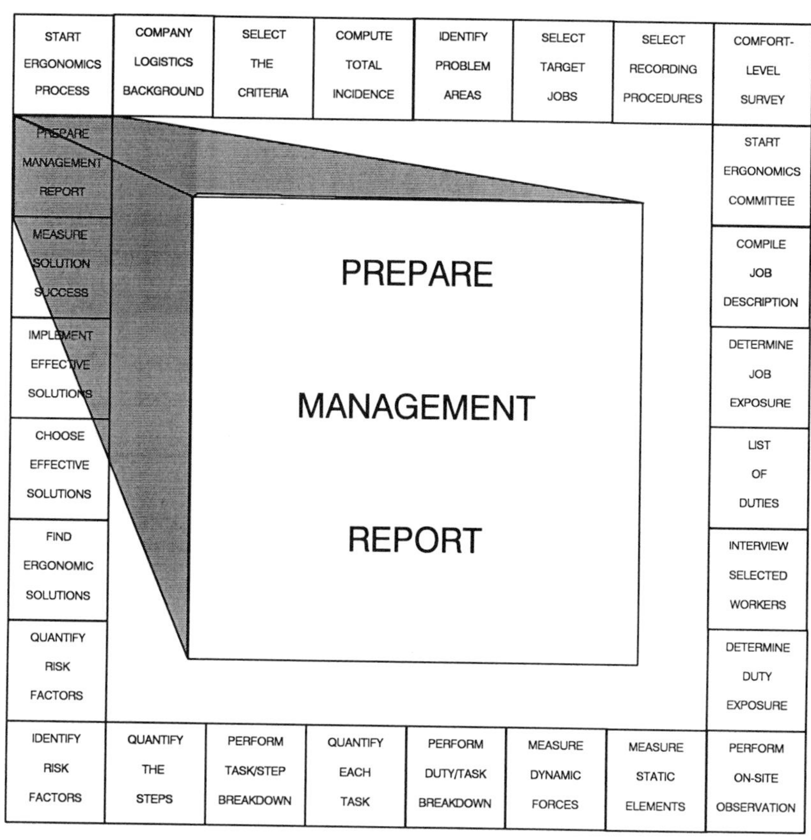
START ERGONOMICS PROCESS
COMPANY LOGISTICS BACKGROUND
SELECT THE CRITERIA
COMPUTE TOTAL INCIDENCE
IDENTIFY PROBLEM AREAS
SELECT TARGET JOBS
SELECT RECORDING PROCEDURES
COMFORT-LEVEL SURVEY
START ERGONOMICS COMMITTEE
COMPILE JOB DESCRIPTION
DETERMINE JOB EXPOSURE
LIST OF DUTIES
INTERVIEW SELECTED WORKERS
DETERMINE DUTY EXPOSURE
PERFORM ON-SITE OBSERVATION
MEASURE STATIC ELEMENTS
MEASURE DYNAMIC FORCES
PERFORM DUTY/TASK BREAKDOWN
QUANTIFY EACH TASK
PERFORM TASK/STEP BREAKDOWN
QUANTIFY THE STEPS
IDENTIFY RISK FACTORS
QUANTIFY RISK FACTORS
FIND ERGONOMIC SOLUTIONS
CHOOSE EFFECTIVE SOLUTIONS
IMPLEMENT EFFECTIVE SOLUTIONS
MEASURE SOLUTION SUCCESS
PREPARE MANAGEMENT REPORT
PREPARE MANAGEMENT REPORT

Prepare Management Report

OBJECTIVE

To generate a short report to management to accurately reflect on-going ergonomics progress.

GUIDELINES

The management of an organization has limited time to devote to any single aspect of day-to-day operation. Therefore it is imperative that reports be as short and concise as possible. They should accurately reflect the efforts of all people involved in the ergonomics process and give credit to key individuals. Essentially, you want to answer the question, "Did it work?"

The fundamental information should be:

1. What was the potential hazard identified?
2. What was done to correct the hazardous condition?
3. Did the intervention work?

Additional information may be provided as an appendix or addendum if your organization responds best to that format.

The additional data would be such items as

1. What criteria were used to determine the workstations to be analyzed?
2. When was the analysis performed?
3. What were the possible interventions discussed?
4. Which intervention was chosen?
5. What was the rationale for choosing that particular intervention?
6. What was the outcome of the intervention?

Be as specific as possible regarding the outcome. If at all possible, put the results into a dollar amount. This will help the management team to assess the overall effectiveness and pass that information along to their superiors. Additional information regarding secondary benefits may also be included. Those secondary benefits should also be as empirical as possible. Even something as simple as a percent difference in tardiness, absenteeism, productivity, or reworks will be more explicit than a general subjective assessment.

Finally, you may choose to discuss future plans to continue to make improvements.

Appendix A: Corporate Ergonomics Program Implementation

A CORPORATE ERGONOMICS PROCESS PROPOSAL

More and more industries are beginning to recognize that they can no longer afford to ignore the growing costs and incidence of overexertion, overuse, or cumulative trauma injuries. This is due to the rising litigious atmosphere, the increasing costs for workers' compensation, the cost of health care, and the increasing responsibility that has been assigned to industry.

In addition, the recent focus of OSHA upon ergonomics has created a sudden urgency to take steps to decrease the chances of unintentional injury to workers. Even when the responsibility is an error of omission, companies and corporations are being cited for failure to provide safe work places.

The following proposal can be used to set the structure of a full-scale corporate ergonomics process. As with everything else presented in this book, take from this plan what is appropriate for your needs. Use it as the foundation for your own unique program. This program should meet the anticipated OSHA standards expected to be released in the near future.

One of the first considerations to ponder is the role that outside consultants are to play in your on-going ergonomic process. Also, you will want to determine how the employees are to be involved. What role will an ergonomics committee play?

Personal accountability is important for the person or persons who are identifying risks and coming up with interventions. Very few, if any, outside consultants will guarantee a reduction in injuries. The level of accountability for an in-house ergonomist or ergonomics committee forces them to be practical and realistic.

On the other hand, an in-house ergonomist may be perceived as a threat on a local level. If on-line management is concerned about disclosure of information to the ergonomist, it can hamper the effectiveness of the ergonomics analysis.

By designing and establishing an in-house ergonomics program utilizing hourly workers functioning in an ergonomics committee, your company can be on the forefront of the injury-reduction effort. It will be prepared for the anticipated OSHA requirements. In addition, the effective implementation of this process will result in significant injury-reduction cost savings.

This type of program utilizes the hourly workers in a gradually evolving role in the ergonomics process. It would be ideal to have the employees involved in the entire process from the very beginning. However, the investment in time away from the line and for initial training and job-site analysis would probably exceed that which is practical. In addition, even if an ergonomics committee were to meet for $1^1/_2$ h every 2 weeks, at the end of the year they would only have put in about 39 h. It is unrealistic to expect an inexperienced group to perform all the risk identification, analysis, and intervention discovery in that limited time.

A more realistic approach is to have workers initially involved in helping to discover interventions to reduce or alleviate unnecessary stresses at the work site. They have an unparalleled expertise in their jobs. They have an intuitive knowledge of what changes will and will not work in their workstation. Given a moderate amount of training, they are capable of working together as a part of an ergonomics committee with a local ergonomics coordinator and management to arrive at effective and reasonable interventions to lessen their chances for injury.

As the program evolves, the hourly workers may be able to perform the risk identification step in the ergonomics process. Once they are able to perform their own risk identification, they will be able to recognize hazards earlier than a more formalized annual or semiannual survey process. This step will necessitate training in risk identification but will not necessitate a comprehensive ergonomics background.

Hourly workers may become a responsible part of the screening, implementation, and tracking phase of job-site analysis process. It is anticipated that after a few years there can be an ergonomics injury-reduction analysis and reduction program running independently in each of your facilities.

Advantages of an Ergonomics Committee

1. Use of their intuitive knowledge: they already know what will probably work or not work
2. Realistic intervention
3. Increased morale, feeling of having some control over their lives
4. Increased compliance with interventions; less reluctance to comply with an intervention simply because it is new and an outsider suggested it

Disadvantages of an Ergonomics Committee

1. Time away from work for training and meetings
2. Improbability that workers will consider interventions that would put their jobs in jeopardy
3. Resentment possible from those not on the committee

One way to evolve into this process is to set up a modest structure of ergonomics coordinators.

GENERAL ERGONOMICS PROGRAM COORDINATOR

This person would either be brought on as a consultant or hired as an employee. The responsibilities would be to:

Write a corporate ergonomics manual containing policies, procedures, and necessary support information

Orient facility staff to the ergonomics process

Train the division ergonomics coordinators

Train the facility ergonomics coordinators

Work with the ergonomics committees on a monthly basis to assist in their function and development

Work with facility and division human resources or safety people to set up effective tracking and reporting mechanisms

Meet with vendors of new ergonomics products and services to assess research validity and value

Act as resource to assess effectiveness of treatment protocols

DIVISION ERGONOMICS COORDINATORS

These individuals will be current safety and health or human resources personnel. They will more than likely have responsibility for several different facilities. They will be trained in a 2- to 3-day program to prepare for this position. Their responsibilities would be to:

Train the ergonomics committee members

Facilitate the information-gathering step of the ergonomic process

Act as resource for facility ergonomics coordinators

FACILITY ERGONOMICS COORDINATORS

These individuals are located in the facilities. They may already have safety and health responsibility, but that is not necessary. Industrial engineers can serve this role as well. The responsibilities would be to:

Perform initial steps of ergonomics process

Meet with and coordinate the intervention discovery step with the ergonomics committee

Present proposals for interventions to facility management

Assist in tracking and assessment of interventions implemented

ERGONOMICS COMMITTEE MEMBERS

These individuals would come from the existing work force at the site or area. They should be made up of combined worker and management groups.

A typical membership would be four or five employees from different areas and representatives from quality, production, and engineering management. The facility ergonomics coordinator would act as the facilitator for the committee activities. The responsibilities would be to:

Discuss potential areas to be targeted for ergonomics analysis

Participate in the discovery intervention process

Assist in determining the most effective interventions

Assist in determining methods for co-worker orientation to the interventions to be implemented

HOW IT WORKS

A facility-based ergonomics committee may be used to work with a company-dedicated ergonomist. The company-dedicated ergonomist can work with local management to determine the criteria for target job selection. For the next step, the workers can provide the background information about specific jobs. Then a comprehensive job-site analysis is performed by the ergonomist to determine specific risk factors. The ergonomics committee can then meet to assist in the intervention discovery, assessment, and screening. Local management can make decisions regarding which interventions to implement and criteria to be used to track the effectiveness of those interventions.

HOW TO GET THERE

Now that you have an idea of the mechanism you would like to be installed in each facility, the task of initiating this process begins. It is going to involve getting a commitment from the upper management level and utilizing that commitment to foster enthusiasm on the facility level. The following steps will help you to achieve that goal. The steps should be performed in the order in which they are presented.

The Corporate Process Outline

1. Adopt plan for process
2. Orient upper management
3. Acquire a statement of commitment
4. Orient facilities/plants
5. Initiate plan (select committee members)
6. Train ergonomics committee
7. Initiate job-site analysis
8. Implement interventions
9. Track effectiveness and report to upper management

Adopt Plan for Process

On a corporate level, it is imperative to form an ergonomics task force. They will share the work necessary to set up this process. They will increase the validity of the proposal to upper management. Also, a group effort will increase input and will help design a realistic and practical approach. This task force will most likely be made up of health and safety professionals. In most cases, the industrial engineers, quality assurance personnel, and production managers have not been made to feel that ergonomics is a part of their responsibility.

ERGONOMIC TASK FORCE ACTIVITY

I. Determine the plan for the corporate ergonomics process
   A. Will the ergonomics committee (E.C.) model be used?
      1. What will be the role of the E.C.?
         All phases of the job-site analysis process
         Risk identification only
         Intervention only
         Other combined phases
      2. How often should the E.C. meet?
   B. Will an ergonomics program director be necessary for this process?
      1. What are the responsibilities of that person?
      2. From where will that person come?
         In-house
         Outside source

II. Establish criteria to determine where to initiate the process
   A. How can current reporting practices be utilized?
   B. How can information already gathered be utilized?
   C. What additional information may be necessary?
   D. What will the reporting responsibilities be?

III. Plan the upper management orientation
   A. Who should present to upper management?
   B. How to use the information already gathered?
   C. What format should it take?
   D. What should come out of that orientation?
   E. What should the commitment contain?

IV. Determine method of rollout to facilities
   A. Who will be responsible for rollout to facilities?
   B. Where to start?
      1. Which plants?
      2. In what order?
      3. To begin when?
   C. At what level?
      1. Corporate wide?

2. Pilot division?
3. Various pilot facilities?

D. Set up time table

V. Determine how to best use information gathered since last meeting

A. How can it be used to assist in getting management commitment?

B. How can it be used to help set criteria for target selection?

C. How can it be used as a measure of the effectiveness of the overall program?

Orient Upper Management

The upper management orientation can be done by representatives of the ergonomics task force, corporate or division health and safety persons, an outside consultant, or an in-house ergonomics expert. This orientation may take the form of a personal meeting, a written proposal, or a series of memos.

Regardless of the personnel involved or the methods utilized, the goals are the same:

1. To provide upper management with information regarding the need for a corporate ergonomics process
2. To explain the ergonomics process
3. To explain the necessity of upper management commitment to ensure success

The following is a proposed agenda for such a meeting.

AGENDA FOR UPPER MANAGEMENT ORIENTATION

A. Presentation of plan
1. Adopt plan for process
2. Orient upper management
3. Acquire a statement of commitment
4. Orient facilities/plants
5. Initiate plan (select committee members)
6. Train ergonomics committee
7. Initiate job-site analysis
8. Implement interventions
9. Track effectiveness and report to upper management

B. Reasons to adopt this plan
1. Cost of doing nothing is exceeding the cost of initiating a plan
    a. Cost of doing nothing (projections)
    b. Cost of the plan
    c. Return on investment of the plan (projections)
2. OSHA mandates a plan

C. Need for commitment on the part of upper management
D. Proposal for upper management commitment

Acquire a Statement of Commitment

The goal of this step is to acquire a written statement of commitment to include an endorsement of the process at each facility and the expected expense to be invested at each facility in the implementation of the interventions. One way to achieve this is to provide a sample statement and allow upper management to customize it to suit their intentions. (See sample "Statement of Management Commitment.") If getting a corporate endorsement is not possible, then the same commitment can be sought on the facility management level.

This statement will convey to all the company managers and personnel that they are authorized and expected to earnestly pursue the development and implementation of an ergonomics process. Secondly, it will state the level of commitment that is expected on the facility level. Thirdly, it will state that implementation of valid and reasonable interventions suggested by the ergonomics committee is expected. Lastly, it will put realistic and firm time frames on the implementation of the local ergonomics efforts.

"IDEAL" STATEMENT OF MANAGEMENT COMMITMENT

I, ______________________________, currently holding the position of ____________________________ am committed to providing the safest environment for all the employees within this organization.

To achieve that goal I hereby state my support of the efforts of the ergonomics committees to achieve that goal.

I hereby authorize the formation of an ergonomics process, which will include management at all levels and workers.

I authorize the participation of employees and managers to meet as a committee and perform job-site analysis and to provide input regarding the need for changes and ways to meet those needs.

This may occur as often as is deemed necessary by the local plant managers but must meet the following minimum requirements:

Selection of committee members	4 hours
Education of committee members	8 hours
Committee meeting	$1^1/_2$ hours per month
Job-site analysis	4 hours per month

Each facility is to commit to the implementation of suggested interventions an amount equal to 10% of the previous year's workers' compensation costs.

I am to be informed of the progress of these groups and the success of the interventions.

Orient Facilities/Plants

The goal of this step is to ensure maximum cooperation with local facility managers. One way to achieve this is to start with written communication from the ergonomics task force or general ergonomics program coordinator to plant managers, on-line managers, and employees. This is then to be followed up with an on-site orientation. If this is impractical, you might consider producing a videotape with a statement by upper management and an outline of the process. The following steps may be helpful in planning your activities:

FACILITY ROLLOUT PROCESS

I. Contact facility manager
II. Meet with facility manager and staff
  A. Agenda for facility management and staff orientation
    1. Outline the process
    2. Outline level of commitment from upper management
    3. Outline what it will cost their facility
    4. Discuss effects on productivity time
    5. Discuss committee meeting frequency
    6. Outline possible equipment needs (video)
    7. Outline their reporting responsibilities
    8. Establish time table
    9. Determine criteria for ergonomics committee selection
    10. Determine method for committee member selection
III. Schedule time to orient on-line management to the process
IV. Orient the on-line management
  A. Agenda for on-line management and team leads
    1. Outline the process
    2. Outline level of commitment by upper management
    3. Outline what it will cost their area
    4. Productivity time considerations
    5. Outline their reporting responsibilities
    6. Review criteria for ergonomics committee selection
    7. Determine method of orienting the employees
V. Orient the employees
  A. Determine the most practical format:
    1. Video?
    2. Posters?
    3. Speaker?
  B. Agenda for on-line employee orientation
    1. Goal of ergonomics process
    2. Request for feedback and input
    3. Need for ergonomics committee members
    4. Methods of selection

5. Call for volunteers
6. How the eventual initiation of interventions can affect them

Initiate Plan (Select Committee Members)

Goal: Set up local ergonomics committees.
Method: Meet with local management to determine best method for ergonomics committee selection and implement that process.
Options: Utilize an outside consultant or ergonomics expert working independent of any worker committees.

The ergonomics committee will work best if kept to about eight members. It is critical that the individuals chosen to participate in the committee understand the importance of their commitment. This may require additional uncompensated time on their part. It should be firmly understood that this is not a way to get out of work.

Ideally, the committee should be made up of either the same number or slightly more hourly workers as non-hourly workers. The ideal profile of a committee member is someone who is imaginative, dedicated, and hard working. Finding such individuals is truly the challenge of this step. Often individuals with a history of a musculoskeletal injury make good participants. A healthy mix of typically vocal and typically quiet workers will help ensure homogeneous employee representation.

It is helpful to have representatives from quality assurance, production, medical services, and industrial engineering. A least one person should have a comprehensive background in ergonomics. One person should take on the role of the facilitator to help keep on track. In most cases, members will take on several roles. For example, the medical person will act as the facilitator, while one of the engineers may act as an ergonomics expert.

AGENDA FOR INITIAL ERGONOMICS COMMITTEE MEETING

1. Establish mission statement
2. Establish level of commitment expected of members
3. Basic orientation to the ergonomics process
4. Brainstorming practice
5. Set time for training

Train Ergonomics Committee

The goal is to provide the ergonomics committee with a clear understanding of their mission, their responsibilities, and the necessary ergonomics background to carry out that mission. This would be achieved through on-site training of ergonomics committee members. If this is impractical or impossible, videotape training programs or off-site ergonomics training may be necessary.

Individuals who complete this course should be qualified to act as a local ergonomics team members. They should be able to perform job-site analyses and develop practical solutions to reduce chances of developing overexertion, overuse, or cumulative trauma injuries.

The program should concentrate on problem solving and brainstorming, rather than a generic checklist approach. Individuals should learn to go beyond the mere identification of a problem and actually think through the multiple options available to decrease the effects of that problem. In addition, they should learn to consider the balance between comfort and safety, and worker productivity and performance. Participants should be able to immediately put to use techniques presented in the training.

A sample course outline would include:

1. Working as a team
2. Brainstorming drills
3. Goals of the class
4. Setting criteria
5. Risk identification
6. Prioritizing
7. Interventions
8. Balance
9. Implementation
10. Tracking

Upon completion of this program, participants should be able to

Identify various sources to establish need for ergonomic changes

Perform a risk factor identification survey

Prioritize identified risks

Develop a comprehensive list of interventions

Weigh the productivity impact of those interventions

Establish an action plan for the implementation of an ergonomic intervention

Track the effectiveness of an intervention

Initiate Job-Site Analysis

The goal of this process is to develop a list of practical interventions aimed at decreasing the unnecessary stresses in the workplace. One way to facilitate this would be to have the facility ergonomics coordinator perform the initial steps of job-site analysis and then present the identified risk factors to the facility ergonomics committee. The committee could then compile a list of possible interventions.

Alternatives to this arrangement would be:

1. Division ergonomics coordinator performs all steps of job-site analysis independently.
2. General ergonomics coordinator performs all steps of job-site analysis independently.
3. Ergonomics committee performs all functions of job-site analysis independently.
4. Combinations of the above.

The steps of the job-site analysis have been discussed in detail in the previous chapters. The following is a restatement of those steps.

Setting criteria. The goal of setting criteria is to determine which areas in a facility will be targeted for analysis and intervention. Specific criteria, such as injured days off or injury incidence rates, are selected and utilized to make this decision. In many cases additional or completely different criteria will be used to measure the effectiveness of an intervention. This alternate or additional criteria may be turnover rates, productivity variances, or subjective comfort-level survey results.

Gathering background. Comprehensive background on each targeted area is collected. This includes identification of the input and output of the operation, a narrative description of the job, and a breakdown of duties, tasks, and, in some cases, steps and components of the job. Input is gathered from human resources, management, and hourly workers. Objective measurement of workstation distances and heights, and static and dynamic forces, is also performed. The targeted area is observed, and in many cases, videotaped.

Risk identification. The videotape is reviewed. Additional information is gathered from the worker whenever appropriate. Using an anatomical region approach, the postural or position stresses, the force exposure stresses, the repetitive stresses, and the environmental exposures are identified and listed. Each risk factor is associated with the corresponding task, step, or component of the job. The result is a comprehensive list of risk factors associated with each part of the job.

Discovering interventions. Each risk factor is considered for reduction utilizing four different approaches. Changes in the input or materials, changes in the output or product, changes in the machine or environment, or actions dealing with the worker are considered and listed. No attempt to screen the interventions is made during this process. It is meant to be a brainstorming approach, resulting in a list of interventions from the most practical to the most farfetched.

Screening interventions. A specific approach is determined and each intervention is considered for implementation. Factors such as cost, return on investment, projected effectiveness, and relative effects on productivity and performance are considered. The most appropriate intervention is chosen.

Implementing interventions. The first step is to let the individuals who are going to be affected by this intervention know that a decision has been made. The next step is to orient that group, being sure to cover why this particular intervention was chosen, why they were chosen, what the expected impact is, and who they can contact should any unforeseen conditions result. The goal is to maximize compliance and, ideally, to foster enthusiasm for the project.

Tracking effectiveness. The criteria for the tracking of effectiveness was already selected. By this time the individuals responsible for compiling this information and preparing a comparative report have also been determined. Any changes in the previously agreed-upon format of the report or the time frames should be settled. The last step will be to assess the effectiveness and to decide whether to expand, amend, alter, or abandon this particular intervention.

Implement Interventions

The goal is the successful implementation of the ergonomics process. It is critical that facilities keep to their commitments. In addition, the ergonomics committees need to be supported with updated information regarding the progress of specific interventions that are initiated throughout the corporation.

Training should be on-going. The division ergonomics coordinators and the general ergonomics coordinator must be available.

In addition, monthly minutes for each ergonomics committee meeting should be forwarded to the general ergonomics coordinator. The information should be the steps of the job-site analysis that have been produced, the efficacy of the criteria-reporting procedures, any breakdowns in the ergonomics processes, the suggested interventions that have been presented to management, and their level of implementation. On-going effectiveness tracking mechanisms should be made a priority.

Track Effectiveness and Report to Upper Management

The goal is to develop an objective measure of effectiveness, both for the overall ergonomics process and the degree of success enjoyed by individual facility interventions. The level of success for each facility and intervention should be communicated both to upper management and to all facilities taking part in the program. This will help facilities to screen their interventions. This would be facilitated by utilizing the following procedures:

1. Local ergonomics coordinator collects information from predetermined sources.
2. Information is sent to division ergonomics coordinators on a quarterly basis.
3. Quarterly reports are prepared by division ergonomics coordinators and copies are sent to the general ergonomics coordinator.
4. Annual report is completed by general ergonomics coordinator.
5. Annual reports are presented to upper management.
6. Information regarding success or failure of specific interventions is communicated throughout the divisions.

Appendix B: Definitions

Duty — A general statement that describes an activity. Examples: loading trucks, cleaning up an area, performing paperwork.

Task — A specific activity that comprises a duty. Examples: sweep floor, move boxes, type letters.

Step — A specific action that makes up part of a task. This will usually start with an action word such as *lift, push, pull,* or *hold.* Examples: lifts bucket, pulls on rope.

Component — A specific anatomical motion at a joint in the body that makes up part of a step. Examples: bends arm up (flexes upper arm), turns wrist to side (ulnarly deviates wrist).

Exposure — The amount of time in hours (or fractions of hours) that a job, duty, task, step, or component is performed each day.

Rate — The rate that a task, step, or component is performed, as expressed in repetitions per hour.

Quantity — The number of times that a task, step, or component is performed.

Task Cycle — A series of tasks that are repeated in a sequence.

Appendix C: Background Information for the Ergonomics Committee

1. List of duties
2. Exposure in hours per day (or percent of time)
3. List of tasks for the duties to be studied
4. Task exposure in hours per day
5. Rate of task performance
6. Total quantities of tasks performed
7. Videotape of tasks in question
8. Map of area
9. Drawing of workstation

Appendix D: Worker Intervention Considerations

EDUCATION PROGRAMS

When considering an education program, try to be as specific as possible about the content. Ask the question, "How could you initiate an education program that would cover . . ."

Less stressful performance methods

Criteria for tool selection

Ways to decrease risk at home

Adjustment of workstation

Any other applicable information

The goal of an education program is to try to bring about some type of adherence to a particular behavior. For workers, that behavior may be lifting in a certain way, periodically stretching, or even pacing work in a certain manner. An education program may also be used to teach management the most useful response to injuries, risk assessment, or job redesign principles.

The necessary objectives for a successful education program are that

1. Participants recognize that the issue at hand really affects them personally. If they do not believe that they may develop cumulative trauma, they are not likely to pay much attention.
2. Participants agree that the behavior being presented will remedy the issue or decrease the risk. If an education program succeeds in the first objective but fails to provide a believable remedy, the results can be very damaging.
3. Participants are given the opportunity to demonstrate the ability to perform the behavior or skill. The participants do not need to be experts, but each one should have performed the behavior or skill correctly at least once. This will facilitate remembering how to perform it at the workstation.

The achievement of these objectives can be enhanced by adhering to four guidelines:

1. Make it job specific. This will help to achieve the first two objectives above. This may be as simple as taking a few slides that demonstrate the undesired behavior at the same facility. An example would be a slide of a worker twisting while lifting.

2. There should be time for interaction. This will allow the presenter to establish his credibility and answer specific challenges regarding the behaviors or procedures presented. In many cases, there are going to be certain vocal individuals who will disagree and argue regarding the validity of the information presented. Those individuals typically will not benefit from an education program. When handled correctly, they do provide a method for convincing the less vocal participants.
3. It is important to provide practice time under the supervision of the presenter to practice and learn the skills presented. Whenever possible this should take the format of both group and individual participation. A good program will focus on the positive aspects of the skills demonstrated by the participants.
4. The training should be entertaining. A dynamic, creative, and fun presentation style is necessary just to maintain attention. An entertaining program will have the participants talking about the program long after it is completed.

Education programs can take many mediums. The most effective is a live interactive presentation. This facilitates achieving all the above objectives. The drawback, of course, is that it involves placing a large number of workers in a room at the same time. This investment in productivity time can be an obstacle to implementing a program. Also, this may limit the number of persons who can attend.

Other forms that allow for less interaction and supervised practice time are interactive computer programs, written literature, and posters. It is important to consider the limitations that may occur due to illiteracy or language barriers. Videotapes, slide shows, or interactive live presentations may be presented by an in-house trainer. That trainer does not have to be a member of the management team. Workers can often prove to be effective trainers when given the appropriate tools and skills. In some cases, it is helpful to follow the program with a quiz.

Worker Conditioning Programs

The goal of a worker conditioning or exercise program is to help the worker better meet the physical demands of the job. This is achieved by making it more convenient for the worker to perform strengthening, flexibility, or cardiovascular exercises. These programs may be aimed at strengthening particular muscle groups to aid in the performance of the job, warm-up exercises, or overall cardiovascular conditioning.

Worker conditioning may take several forms, such as group stretching exercises at the start of every shift, having exercise equipment available at the work site, or subsidizing a worker's membership in a health club.

Personal Protective Equipment

The kind of personal protective equipment that may be considered is mostly in the form of the braces, splints, and belts. The goal of braces or splints is to limit the exposure to a potential hazard. While these have been used as postinjury procedure, many companies are attempting to utilize these as a preventative device. There are basically two methods. One is to restrict motion and the other is to provide support. An example of the first is a wrist splint, which restricts the wearer to a neutral position. A concern is, first, that the restriction of motion at the wrist may result in the stress being moved to the elbow, shoulder, or upper back. Also, by restricting the motion there is less natural muscle action in the body part. That natural action facilitates circulation and fluid motion in the joints necessary to keep them healthy.

An example of the "support" type of device is a lift belt. The amount of actual support provided by the belt is questionable. In addition, there has been no scientific evidence that any one type of belt works any better than any other type of belt. Therefore it is possible that the belt itself is not as important in providing support as it is in reminding individuals to lift in a certain manner. This type of program should not be considered a substitute for redesigning an unsafe workstation or providing training in the necessary skills.

Monitoring Devices

The goal of a monitoring program is to gather or provide information regarding the current status of a worker's performance or his body's response. This may be as simple as attaching a device to a worker's back to provide a signal when he is lifting in an incorrect manner. This type of instantaneous feedback is meant to facilitate adherence to a particular lifting technique.

Another type of monitoring is performed over time. Individuals may be tested at a point in time to establish a baseline. They will then be periodically retested to assess any changes. This may take the format of a physical fitness test to measure strength, flexibility, and endurance. It may also be more anatomically specific. Devices such as a neurometer or vibrometer may be utilized to assess the integrity of nerves within the body. Any changes from the baseline can signal a need for special exercises, job redesign, or more specific testing and treatment.

Job Rotation

The goal of job rotation is to limit exposure to a potential ergonomic hazard. This has been effective to varying degrees, depending more on the individual situations. In some cases, it can work very well. In other cases, there are some drawbacks. A rotation program should be set up by a qualified individual with a solid biomechanical background. You want to be certain that individuals are not using the same muscles as they rotate from one task to another. In addition, many rotation programs assume an equal level of skill. The ideal rotation program would potentially have warehouse workers loading trucks one day and working in the office typing the next day; however, it is unlikely that the warehouse workers would possess the typing skill necessary to work in the office. Lastly, to implement a rotation program may upset a long established social order. Individuals who may have worked for an employer for a long time will often gain the seniority to hold the less stressful jobs. It can be disruptive to suddenly change the rules and ask them to return to performing the less desirable tasks.

Worker Selection

The goal of the worker selection process is to match the worker with the job. This may involve assessing an individual's skills, grading his or her bias for accidents, listing factors that may predispose the individual to an injury, or a physical testing component. The physical testing component may include strength, flexibility, or aerobic fitness in order to verify that an individual is suitable for the job.

This should be thought of as a way of protecting both the worker and the employer. The worker does not want to be injured. At the same time, the employer does not want to have a worker injured.

Worker selection may take the form of a subjective assessment on the part of the personnel manager, a physical examination performed by a physician, a comprehensive medical history, a chest x-ray, a psychological profile questionnaire, a localized electromyographic test, an isokinetic back-strength test, or a sub max VO_2 test.

It is critical to examine all the legal ramifications of utilizing various techniques for worker selection. Laws and guidelines, such as those presented by the EEOC, the Rehabilitation Act of 1973, the Civil Rights Act, and the Americans with Disabilities Act, need to be carefully examined.

Appendix E: NIOSH Formula

$$\text{Action Limit} = 90(6/H)(1 - 0.1\,|V-30|)(.7 + 3/D)(1 - F/F_{max}),$$

where H = horizontal distance of the hand from the the body's center of gravity at the beginning of the lift, V = vertical distance from the hands to the floor at the beginning of the lift, D = distance that the object is lifted vertically, and F = average number of lifts per minute.

Appendix F: Additional Readings

Alexander, D.C., *The Practice and Management of Industrial Ergonomics,* Prentice-Hall, Englewood Cliffs, NJ, 1986.

Alexander, D.C. and Pulat, B.M., *Industrial Ergonomics: A Practitioner's Guide,* Industrial Engineering and Management Press, Norcross, GA, 1985.

American Conference of Governmental Industrial Hygienists, *Ergonomic Interventions to Prevent Musculoskeletal Injuries in Industries,* Lewis Publishers, Chelsea, MI, 1987.

Astrand, P.O. and Rodahl, K., *Textbook of Work Physiology,* McGraw-Hill, New York, 1977.

Basmajian, J.V., *Primary Anatomy,* Williams & Wilkins, Baltimore, 1976.

Brammer, A.J., *Vibration Effects of the Hand and Arm in Industry,* Wiley and Sons, New York, 1982.

Burgess, J.H., *Designing for Humans: The Human Factors in Engineering,* Petrocelli Books, 1986.

Chaffin, D.B. and Anderson, G.B.J., *Occupational Biomechanics,* John Wiley and Sons, New York, 1984.

Cook, J.D., Hepworth, S.J., Wall, T.D., and Warr, P.D., *The Experience of Work,* Academic Press, New York.

Eastman Kodak Company, *Ergonomic Design for People at Work,* Van Nostrand Reinhold, New York, 1983.

Grandjean, E., *Ergonomics and Health in Modern Offices,* Taylor and Francis, London, 1983.

Grandjean, E., *Fitting the Task to the Man: An Ergonomic Approach,* Taylor and Francis, London, 1980.

Heyward, V.H., *Design for Fitness,* Burgess Publishing, Edina, MN.

Joyce, M. and Wallersteiner, U., *Ergonomics: Humaninzing the Automated Office,* South-Western Publishing, Cincinnati, OH, 1989.

Kendall, F.P. and McCreary, E.K., *Muscles Testing and Function,* Williams and Wilkins, Baltimore, 1983.

Landy, F.J. and Farr, J.L., *Measurement of Work Performance,* Academic Press, New York, 1985.

McCormick, E., *Human Factors Engineering & Design,* McGraw-Hill, New York, 1976.

N.I.O.S.H., *A Work Practices Guide for Manual Lifting,* Department of Health and Human Services, Cincinnati, OH, 1981.

Perkins, R., *Outline of Gross Anatomy,* Northwestern University Medical School, Chicago, IL, 1981.

Rodgers, S.H., *Working with Backache,* Perinton Press, Fairport, NY, 1985.

Rowe, M.L., *Backache at Work,* Perinton Press, Fairport, NY, 1986.

Rowe, M.L., *Orthopaedic Problems at Work,* Perinton Press, Fairport, NY, 1985.

Tichauer, E.R., *The Biomechanical Basis of Ergonomics,* John Wiley and Sons, New York, 1978.

Zenz, C., *Developments in Occupational Medicine,* Medical Publishing.

Appendix G: Sample Ergonomics Analysis

The following short example is meant as a guide to integrate the work done in an ergonomics analysis. First, an example of a complete report will be provided. Following that example, the format will be outlined.

ERGONOMICS ANALYSIS
A.V.B. INCORPORATED
October 22, 1991

BACKGROUND

On October, 22, 1991, an ergonomics analysis was performed at A.V.B. Inc. located at 753 Industrial Drive, Bronx, New York. The major product produced at this facility is industrial training aids.

A.V.B. Inc. employs 350 persons working two shifts. In the past three years there has been a change in facility management. In addition, lift belts and back injury prevention training have been initiated on a pilot program. No objective data are available regarding the relative success of these interventions.

The job title that was analyzed was the Order Picker and Packer. Seventy-eight persons are employed to perform this job function.

JOB DESCRIPTION

Order picker/packers are responsible for pulling materials from storage shelves, packing the various pieces into boxes, addressing boxes, and transporting them to the postage and shipping area.

Order pickers/packers work 5 days a week. They are given two paid 15-minute coffee breaks and one unpaid 45-minute meal break. The total job exposure is $7^1/_2$ hours per day.

DUTIES LIST AND EXPOSURE

Order Pickers/Packers

Duties	Exposure (hours)	Percent of Day Spent
Retrieving materials	4.00	53%
Packing boxes	1.25	17%
Addressing boxes	1.50	20%
Transporting boxes by cart	0.75	10%
Other	0.00	0%
Total job exposure	7.5	100%

ERGONOMICS ANALYSIS

A.V.B. Incorporated
Order Picker/Packer — Mike Burke
006-34 — 11-15-91

While performing the task of getting orders from stock, which is performed 4 hours a day, at a rate of 20 orders an hour, for a total of 80 orders a day, the subject is exposed to the following risk factors:

1. Assumes an end range position of the ankle — turns foot out.
2. Assumes an end range position of the low back — twists at waist.
3. Assumes an end range position of the shoulder — moves arm overhead in front.
4. Assumes an end range position of the wrist — turns hand over (down or in).
5. Maintains an unsupported position at the low back — prolonged standing.
6. Applies a lifting force of 25 pounds average pounds with the low back.
7. Applies a lifting force of 25 pounds average pounds with the shoulder.
8. Applies a lifting force of 25 pounds average pounds with the elbow.
9. Foot/ankle is exposed to hard or unpadded surfaces (cement floor).

Worker productivity, performance, comfort, safety, and health can be enhanced by the initiation of the following ergonomic interventions:

- Initiate an education program consisting of:
  - activities to do at work and at home to decrease injury risk;
  - exercises to be performed at work; and
  - a list of home activities that can increase injury risk.
- Rotate workers between warehouse work and office work.
- Place only infrequently needed items on higher shelves.
- Place all stock on gravity-fed shelving.
- Provide workers with sorbathane or other shock-absorbing insoles for shoes.
- Have workers wear lift belts to maintain less stressful posture when lifting.
- Mandate specific procedures for lifting and carrying materials and reward compliance with incentive programs.
- Initiate an exercise program for low back flexibility and strength maintenance to be performed at the beginning of each shift.
- Utilize a monitoring device to provide feedback when worker is performing an activity in a stressful way.
- Use a spring-loaded double-bottom or self-adjusting cart or bin.
- Decrease rolling resistance of cart, carriers, or conveyance devices by changing wheels.
- Place ramps over thresholds.

SUMMARY AND RECOMMENDATIONS

The job of Order Picker and Packer was identified by Mr. Smith, the Health and Safety Coordinator, for ergonomic analysis. This was as a result of several recent low back injuries and a shoulder injury. Following this analysis, several controllable risk factors that may affect worker performance, productivity, comfort, and safety were identified.

The primary job duty is retrieving materials from stock. This task was analyzed. The primary risk factors were associated with material handling demands.

Of all the intervention methods mentioned above, the most appropriate would be the initiation of an education program, initiation of a maintenance program to keep carts in the most effective condition, providing cushioned insoles for workers, and reorganizing shelves to put least called-for items on the higher shelves.

FORMAT

ERGONOMICS ANALYSIS
(Company Name)
(Date)

1. Background

On (date), an ergonomics analysis was performed at (company name), (address). The major product produced at this facility is (product).

2. Summary of company background

3. Job description

4. Job background and exposure summary

5. Duties list and exposure

6. Ergonomics analysis

(Company Name)
(Job title) (Analyst)
(Job number) (Date)

While performing the task of (task description), which is performed (task exposure) per day, at a rate of (rate) units per hour, for a total of (quantity) units per day, the subject is exposed to the following risk factors:

(List risk factors.)

These risk factors have been associated with an increased incidence of cumulative trauma disorders. Many of these conditions can be significantly reduced by the initiation of the following ergonomic interventions.

(Ergonomic interventions.)

7. Summary and recommendations

Index